Natural History and Ecology of Mexico and Central America

Edited by Levente Hufnagel

Published in London, United Kingdom

IntechOpen

Supporting open minds since 2005

Natural History and Ecology of Mexico and Central America
http://dx.doi.org/10.5772/intechopen.77889
Edited by Levente Hufnagel

Contributors
Jose Luis Godínez-Ortega, Juan V. Cuatlán-Cortés, Juan M. López-Bautista, Brigitta I. van Tussenbroek, Rebeca Monroy-Torres, Graciela Velazquez Delgado, Erika Carcaño Valencia, Gilber Vela Gutiérrez, Carlos Gonzalez Gandara, Ernesto A. Aarón Chávez, Luis Mejia-Ortiz, Guillermo Ruiz-Cancino, Juan C. Tejeda-Mazariegos, Jair Gaspar Valladarez, Alejandro L Collantes-Chavez-Costa, Oscar Frausto-Martinez, Germán Yáñez, Peter Sprouse, Miguel Ángel Reyes-López, A. Alonso Aguirre, César Paúl Ley-Quiñónez, Alan Alfredo Zavala-Norzagaray, Kevin Alan Zavala-Felix, Valeria Leal-Sepúlveda, Catherine E. Hart, Fátima Yedith Camacho-Sánchez, Laura Yáñez-Espinosa, Alberto Prado, Jacqueline Bede, Gabriel Rubio-Méndez, José Arturo De-Nova, Joel Flores, Juan Antonio Reyes-Agüero, Joaquín Sosa-Ramírez, Vicente Díaz-Nuñez, Diego R. Pérez-Salicrup, Manfredo Turcios, Richard LaVal, Mario Martínez, Hefer Ávila-Palma, Levente Hufnagel, Ferenc Mics

First published in London, United Kingdom, 2021 by IntechOpen
IntechOpen is the global imprint of INTECHOPEN LIMITED, registered in England and Wales, registration number: 11086078, 5 Princes Gate Court, London, SW7 2QJ, United Kingdom
Printed in Croatia

British Library Cataloguing-in-Publication Data
A catalogue record for this book is available from the British Library

Additional hard and PDF copies can be obtained from orders@intechopen.com

Natural History and Ecology of Mexico and Central America
Edited by Levente Hufnagel
p. cm.
Print ISBN 978-1-83968-482-1
Online ISBN 978-1-83968-486-9
eBook (PDF) ISBN 978-1-83968-487-6

For EU product safety concerns:
IN TECH d.o.o., Prolaz Marije Krucifikse Kozulić 3, 51000 Rijeka, Croatia, info@intechopen.com or visit our website at intechopen.com.

Meet the editor

Dr. Levente Hufnagel is an associate professor and the head of the Research Institute of Multidisciplinary Ecotheology, John Wesley Theological College, Budapest, Hungary. He works on ecology, biogeography, ecological research methodology, and sustainability and has more than twenty years of experience leading Hungarian academic institutions. He has taught and supervised Ph.D., MSc, and BSc students from various social and cultural backgrounds. He has published more than 240 scientific publications (on both aquatic and terrestrial ecological aspects of plants, animals, and microbes at the community as well as population levels) and has more than 1000 independent citations. As a participant in several big ecological research and development projects, Dr. Hufnagel has significant experience in multidisciplinary collaborations (with more than 200 coauthors in different publications). He is editor-in-chief of an international scientific journal. Dr. Hufnagel graduated from Eötvös Lorand University, Budapest, Hungary, with a master's degree in Ecology and Evolutionary Biology and a Ph.D. in Hydrobiology. He also has a Ph.D. in Agricultural Science from Szent István University as well as other degrees from the Corvinus University of Budapest and Adventist Theological College.

Contents

Preface

Mexico and Central America is a very special, interesting, and important region from faunistic, floristic, and ecological viewpoints. Mesoamerica (starting from the southern states of Mexico) differs from Central America, which is a more general geopolitical name. The expression "Middle America" is in use as well, which involves all areas south from the border of the United States including the islands of the Caribbean.

In the maintenance and conservation of our planet's biodiversity, knowledge of current biogeographical patterns, Earth-historical changes, speciation, or invasion processes, flora, fauna, natural history, and ecology have outstanding importance. Many areas of biodiversity are still completely undiscovered. The ecological impacts of global climate change, urbanization, overpopulation, environmental pollution, deforestation, land cover, and land-use changes also have a significant impact on biodiversity, flora, and fauna through biogeographical patterns.

This book presents interesting information about a wide spectrum of topics including the biodiversity and ecology of plant communities, tropical subterranean and marine ecosystems, endangered and rediscovered species, and applied or human ecological topics, fishing, urbanization, and sustainable diet.

Natural History and Ecology of Mexico and Central America gives a very interesting report about the frontiers of this geographical area. I am sure that this book will be very useful for everybody who would like to get some insight into the recent problems of biodiversity research and ecology of this region.

Levente Hufnagel
Research Institute of Multidisciplinary Ecotheology,
John Wesley Theological College,
Budapest, Hungary

Section 1

Natural History and Ecosystems

Chapter 1

Introductory Chapter: Biodiversity of Mexico

Levente Hufnagel and Ferenc Mics

1. Introduction

1.1 Flora, fauna and vegetation

Mesoamerica (starting from the southern states of Mexico) differs from Central America, which is a geopolitical name. The expression "Middle America" is in use as well, which involves all areas south from the border of the US including the Islands of the Caribbean [1]. In this chapter the biodiversity of Mexico is presented and the literature on its protection is analyzed.

Mexico being the largest country in the region is very rich in species in itself (**Tables 1–4**). Lot of species occur even in the dry northern areas. In the Chihuahuan Desert 826 plant species are noted by Villarreal-Quintanilla et al. [7], out of which 560 are endemic, 165 are quasi-endemic and és 176 are microendemic. 116 taxa can be originated from a non-arid habitat. The most species-rich are Cactaceae with 141, Asteraceae with 106, Boraginaceae with 34 and Brassicaceae with 31 species. On the California Peninsula 723 endemic species are noted by Riemann and Exequiel [8], claiming that the great number of endemic species is due to the heterogenity of the environment. The flora and fauna are very interesting because the area of Mexico involves the border of *Neotropis* and *Nearctis* (Mexican Transition Zone), which is not exactly a border but rather a wide transition zone, its accurate definition is yet to be created. The determinaion is based on the distribution of the endemic genera characteristic to one or the other area. The several results obtained regarding this vary hugely. The determination of the location and width of the transition zone is different among authors without a consensus, which requires further floristic examinations [9]. Vegetatio varies depending on the topography that has a great role in the fromation of the great number of endemic species as well [10].

Group	Number of species worldwide	Number of species in Mexico	Worldwide/Mexico %	Number of endemic species	Endemic species %
Vascular plants	248 428	18000–30000	7–12	10000–15000	33–50
Amphibians	4222	284	7	169	60
Reptiles	6458	717	11	368	51
Birds	9040	1050	12	125	12
Mammals	4629	450	10	140	31

Table 1.
Biodiversity of Mexico according to Mittermeier et al. [2].

Vegetation	% area	Number of species	Endemisms %
Cloud forest	0.7	3000	30
Rainforest	4.4	5000	5
Pine and oak	12.9	7000	70
Xerophyl and steppe	34.8	6000	60
Wetlands	1.43	1000	15
Tropical deciduous	7.9	6000	40
Agricultural and ruderal	ND	2000	20

Table 2.
Flora of Mexico according to vegetation types based on Flores-Villela and Gerez [3].

Biom/ecoregions	Area (km^2)	Estimated number of species	Remaining intact vegetation %
Tropical and Subtropical Moist Broadleaf Forests			
Chiapas moist forests	5759	3000–4500	67
Chimalpas montane forests	2076	2000–3000	49
Pantanos de Centla	17152	1500–3000	11
Petén-Veracruz moist forests	148604	5000–8000	31
Sierra de los Tuxtlas	3890	2400–3500	11
Sierra Madre de Chiapas moist forests	11218	3500–4700	34
Veracruz moist forests	68949	4500–7000	20
Veracruz montane forests	4942	2200–3500	55
Yucatán moist forests	69485	1300–1900	64
Tropical and Subtropical Dry Broadleaf Forests			
Bajío dry forests	37282	2900–5000	0,64
Balsas dry forests	62249	2500–5100	2
Central American dry forests	67777	2800–400	12
Chiapas Depression dry forests	13974	1500–3500	7
Jalisco dry forests	26051	1000–2500	26
Revillagigedo Islands dry forests	210	ND	ND
Sierra de la Laguna dry forests	3975	500–1000	0,008
Sinaloan dry forests	77364	1700–2500	13
Sonoran-Sinaloan transition subtropical dry forest	50326	ND	ND
Southern Pacific dry forests	42283	2500–5100	15
Veracruz dry forests	6616	900–2000	5
Tropical and Subtropical Coniferous Forests			
Central American pine-oak forests	110948	4000–6000	42
Sierra de la Laguna pine-oak forests	1061	700–1200	4

Biom/ecoregions	Area (km^2)	Estimated number of species	Remaining intact vegetation %
Sierra Juárez and San Pedro Mártir pine-oak forests	4000	ND	ND
Sierra Madre de Oaxaca pine-oak forests	14299	2500–3700	55
Sierra Madre del Sur pine-oak forests	60976	3600–5000	43
Sierra Madre Occidental pine-oak forests	222700	ND	ND
Sierra Madre Oriental pine-oak forests	65600	ND	ND
Trans-Mexican volcanic belt pine-oak forests	91800	4000–6000	26
Tropical and Subtropical Grasslands, Savannas, and Shrublands			
Western Gulf coastal grasslands	77425	2150–2250	3
Deserts and Xeric Shrublands			
Baja California desert	45940	1500–2200	8
Central Mexican matorral	59195	2500–4500	0,011
Chihuahuan desert	501896	3300–3600	50
Gulf of California xeric shrub	22573	900–1900	29
Meseta Central matorral	124975	3000–4500	4
San Lucan xeric scrub	3685	ND	12
Sonoran desert	260000	2600–3000	37
Tamaulipan matorral	16300	1500–2500	9
Tamaulipan mezquital	141500	1700–2500	23
Tehuacan Valley matorral	9842		0
Flooded Grasslands and Savannas			
Central Mexican wetlands	259	100–600	ND
Montane Grasslands and Shrublands			
Zacatonal	306	150–500	ND
Mangroves			
Alvarado mangroves	4534	20–400	1,12
Marismas Nacionales-San Blas mangroves	2034	20–400	
Mayan Corridor mangroves	4079	20–400	
Mexican South Pacific Coast mangroves	1168	20–400	
Petenes mangroves	1971	20–400	
Ría Lagartos mangroves	3457	20–400	
Tehuantepec-El Manchon mangroves	2685	20–400	
Usumacinta mangroves	3118	20–400	
Mediterranean Forests, Woodlands, and Scrub			
California chaparral and woodlands	121000	1550–1750	1

Table 3.
Bioms and ecoregions of Mexico according to Kier et al. [4] and Dinerstein et al. [5].

State	Birds	Mammals
Veracruz	664	101
Chiapas	628	90
Oaxaca	687	116
Jalisco	481	107
Guerrero	476	72
Puebla	367	76
San Luis Potosí	469	93
Michoacán	460	79
Chihuahua	329	95
Edo. México	281	55
Tamaulipas	444	90
Sonora	456	100
Durango	308	81
Nayarit	407	72
Nuevo León	252	63
Hidalgo	236	59
Morelos	274	46
Sinaloa	460	69
Coahuila	275	80
Tabasco	370	47
Baja California	292	95
Quintana Roo	340	51
Distrito Federal	222	44
Yucatán	343	58
Baja California Sur	258	77
Colima	318	51
Zacatecas	154	75
Guanajuato	256	45
Campeche	281	50
Querétaro	181	36
Aguascalientes	89	33
Tlaxcala	89	21

Table 4.
Bird and mammalian fauna of Mexico according to CONABIO [6] (National Commission for knowledge and use of biodiversity) by states. In coastal states marine mammals are included.

The climate formed by the emerging mountains (Sierra Madre Oriental, Sierra Madre Occidental, Trans-Mexico Volcanic Belt) also affects evolutionary processes resulting in the development of new species [11, 12]. The richness of species and the species composition of the community depends on the heterogenity and diversity of the environment [10]. As Moonlight et al. [13] presents on the example of *Begonia* genus regarding the DNA sequence data of the plastis diversification is fast and new species develop in the hetrogeneous environment. According to their results there were two indipendent colonization events from Africa towards the Neotropis. Two different clads were reconstructed, which diversified around the middle of miocene in South America and radiation occurred once towards Central America and Mexico.

Today tropical deciduous forest is typical on the western side but it is fragmented due to human activity and only 3% is protected. Agriculutal areas are concentrated in areas where the climate is seasonal that is why the reduction is faster than in the case of rainforests [14]. 10% and 19.7% of tropical forest and shrubland, respecively are protected. On the east side 28% of the rainforests are protected (Estado [15]). Examinations of flora and fauna are still not complete, our knowledge on the wildlife of the area is deficient [16, 17]. New species are still identified in Mexico such as *Tryonia* (Caenogastropoda: Cochliopidae) species discovered by Hershler et al. [18] in the creeks of the Chihuahuan Desert (Durango State). Some of these probably have become extinct since they were failed to be found again. The cause of their extinction is the destruction and disappearance of their habitat. According to forcasts the average annual temperature will rise by 3,7–3,8°C to 2090 in Mexico, the amount of precipitation will reduce by 18,2% and AAI (Annual Aridity Index) will rise by 26% [19]. These will result in a 25% decrease in the value of ENS (effective number of species), if the forcasts prove to be accurate [20]. Characteristic plants of Mexico the cacti (Cactaceae) can also become rare by then due to land transforming activity of man, 31% of the species are already endangered. Climate change may cause other species to become endangered. Cacti are drought tolerant but because of climate change species may become endangered due to the reduction or extinction of pollinators and animals dispersing seeds. In the case of cacti often ants carry the seeds (Myrmecochory) only to limited distances. Many species have a small area of distribution (75–100 km^2) as well, making them particularly vulnerable [21]. A remarkable example of the destruction of human activity to habitat is the wall that is being built on the border of the US and Mexico, which parts that have already been set up initiated the reduction of size, quality and connections of habitat in the otherwise diverse area [22]. Migration of many amphibian, reptile and mammal species are hindered by the barrier and human activity. This becomes relevant when certain species look for new habitat due to climate change. Obstructing their movement endanger species or certain popoulations. The genetic diversity of species reduce with the extinction of populations endangering the entire species [23].

2. Environmental protection and biodiversity conservation

Since the nineties Mexico have been involved in the work of international environmental organizations with increasing activity and now have signed 44 international agreements. It is an active stakeholder of organizations such as the CBD (Convention on Biological Diversity) and CITES (Convention on International Trade in Endangered Species of Wild Fauna and Flora), CEC (Commission for Environmental Cooperation), Trilateral Committee, The Ramsar Convention on Wetlands, EMSA (Mesoamerican Biological Corridor and the Mesoamerican Strategy of Environmental Sustainability). The NDP (National Development Plan) can be considered as a legal framework within the country providing basic regulations for the federal government by giving guidance for the work of the government with the development of clear strategies, setting regional aims and measures to be implemented, coordinating institutional and regional programs involving several areas [24].

The program, in which Mexico also participates, aiming to preserve biodiversity is the Payments for Ecosystem Services (PES) that is an economic framework to plan and introduce payment schemes that provide market remuneration for ecosystem services [25]. This tool seems to be eligible for the protection of rainforests in the developing countries. 2.6 million hectares are involved in the program in

Mexico, which in terms of money means 450 million USD and this is one of the biggest among such programs in the world [26]. With these territories approximately 25% of the biodiversity that needs protection is now included in the program [27]. According to Honey-Rosés et al. [28] 3–16% more forest were managed to be preserved along with those habitats that these forests include. Deforestation in those areas that are involved in the program is carried out at a slower rate, than in those that are not invovled. This can be beneficial to the population and to the owners of the forests as well since it can lead to other income sources, such as with the development of ecotourism. In the protected areas it is also important to invovle local residents and educate them on the importance of wildlife protection since mostly they only experience difficulties in accessing their resources [29, 30].

In order to preserve at least some part of biodiversity a system of protected areas is required. Determination of the most valuable protected areas can be carried out by ecological modeling. Torres-Miranda et al. [31] used red oaks and their distribution area to estimate (section *Lobatae*; 75 species) those area that worth the most concerning protection.These species representing ecosystems prove to be useful indicators for conservation professionals.These species occur in various habitats often as dominant species and have an important part in preserving diversity. Based on a computer simulation (Complementarity analyses) 12 current areas under protection should be enlarged and 26 new should be established with a total of 512 500 ha area to ensure adequate protection for habitats. Certain species, especially arthropods bond to certain plant communities. With the protection of fast reducing oak forests species bonded to them can also be preserved [32].

An experiment in the Tehuacán Valley showed that biodiversity is preserved with greater succes in areas under forest management than in areas that are not involved. Although, preserving rare species is limited even in this system. In average 59% of plant species and 94% of the genetic variety of dominant species (*Polaskia chichipe*, *Escontria chiotilla*, *Myrtillocactus schenckii*) was managed to be preserved in the examined area. In areas that are not involved the remaining natural flora decrease fast due to the increasing intensity of agriculture [33]. The number of species that can be preserved depends on whether the areas in question are private or community properties, also on the availability of natural resources, ecosystem services to people, as well as on the culture of certain communities. Ornamental or herbal species and those that are used for human or animal nutrition tend to be preserved even if the area gets involved in agriculture. Lanes dividing plots and islands within the agricultural area promote later regeneration [34]. Traditional methods of agriculture have less harmful effects on biodiversity than the current intensive mechanized agriculture. In the state of Oaxaca indigenous people have been carrying out agricultural production for centuries, which lead to a mosaic landscape with patches of forest and agricultural land. Nowadays due to urbanization more and more people give up farming and move to cities. On lands left behind a secondary forest forms, which surprisingly lead to the decrease of biodiversity.

3. Agricultural aspects

Traditional agriculture plays a part in sustaining biodiversity, since the landscape renews regularly. Traditional farming also has to be considered in the decision making process related to protected area [35]. Larios et al. [36] also claim that traditional farming has a great role in the preservation of biodiversity. According to a survey carried out in the area of the Tehuacán Valley 281vascular plants were identified even in the gardens out of which 34% were endemic. Though abundance of cultivated plants was the largest. The highest value (199) was found in gardens lying near the cloud forests in the mountains. The lowest valule (141) was found in those that are located near deciduous forests. This can be explained by the tendency that owners cultivate plants in their own land to compensate the rarity of useful species in the nearby forests.

An agricultural effort to protect the diversity of the species is the production of shade-grown coffee [37]. With the production of shade-grown coffee most of biodiversity can be preserved since a proportion of the original vegetation survives. Coffee produced with this method has a high price, which can encourage more and more people to choose this cultivation method. Cultivation area is increasing unfortunately to the detriment of the primary forest, thus in its current form this is not the appropriate method to preserve biodiversity [38]. For the protection of marine ecosystems aquaculture is spreading in Mexico as well regarding both fishes and marine invertebrates. The development of the technology means income for the economy and wildlife can also be protected since the importance of illegal fishing decreases. Aquacultural production can mean a legal income source for the local people, while overfishing and the decrease of biodiversity can be avoided [39].

Phytoremediation plays and important part in the region as well in the nutralisation of industrial pollutants, therefore the research of those organizations are important, which can be used for this purpose [40]. An abandoned mine in the state of Hidalgo was recultivated and the area was reforested. 56 species representing 29 families were managed to be planted. Samples of AMF (Arbuscular Mycorrhizal Fungi) from tree roots *Glomus* (Glomeraceae: Glomerales) and *Acaulospora* (Acaulosporaceae: Diversisporales) species were the most common. These have a great role in phytoremediation due to incereasing the trees tolerance against heavy metals in the extreme environment, as well as decreasing the distribution of toxic subsatnces in the environment [41]. Harmful effects of industrial pollutants can be reduced by phytoremediation, therefore it serves the protection of wildlife. Regarding the big biodiversity of the area it is likely that new species will be found to be suitable for this purpose. There is another example of a microbial scale biodiversity research with direct economic benefits. Diversity of bacteria and fungi living around cultivated *Agave tequilana* roots in the soil, rizosphere and phyllosphere, in the endosphere of the root and the leaf was compared by Coleman-Derr et al. [42] with similar microbes of wild *Agave salmiana* and *Agave deserti* populations. *Agave tequilana* can be cultivated in areas where no other crop can survive and can be used for bioethanol production. Symbiotic microorganisms influence plant health and accomodation to stress due to this the rate of growth as well [43]. Manipulation of microbiom may increase the rate of growth and therefore, the amount of ethanol that can be produced [44]. The composition of the microorganism community based on the analysis of the traceable DNA changes depending on the compartment, which was obtained by the amplification of ITS2 and16S regions. Geographical distributation also affected composition. In the case of cultivated plants alpha diversity was low, which can be explained with agricultural practices.The community is dominated by the genera of Enterobacteriaceae family (*Pantoea*, *Leclercia*, *Trabusiella*), therefore soft rot disease became often, which cost millions. Genetic diversity of plants is also low due to vegetative reproduction. That is why the bacteria *Pantoea agglomerans* could develop avoiding strategies during the evolution against the plants defense mechanisms [42]. Apart from agave, oil pressed from the seed of *Jatropha* (Euphorbiaceae) species, especially from the seed of *J. curcas*, is also appropriate for the production of biodiesel [45]. 50 of the 186 species occur in Mexico as well, most of which are endemic. They are distributed from the rainforests to the deserts everywhere. Their distribution is limited by the frost in the mountains. Other species may be eligible for agricultural use, which require further research [46].

Area, estimated number of species and percentage of remaining intact vegetation. The boundaries of ecoregions are not the same as national borders.

Author details

Levente Hufnagel[1*] and Ferenc Mics[2]

1 Research Institute of Multidisciplinary Ecotheology, John Wesley Theological College, Budapest, Hungary

2 Department of Environmental Security, John Wesley Theological College, Budapest, Hungary

*Address all correspondence to: leventehufnagel@gmail.com

References

[1] Winkler, K. (2011): Middle America, not Mesoamerica, is the accurate term for biogeography. – The Condor 113(1): 5-6.

[2] Mittermeier, R.A., Myers, N., Mittermeier, C.G. (1999): Hotspots: Earth's biologically richest and Most endangered terrestrial ecoregions. – Journal of Mammalogy 83(2): 630-633.

[3] Flores-Villela, O., Gerez, P. (1994): Biodiversidad y conservación en México: vertebrados, vegetación y uso del suelo. – Comisión Nacional para el Conocimiento y Uso de La Biodiversidad y Universidad Nacional Autónoma de México, México.

[4] Kier, G., Mutke, J., Dinerstein, E., Ricketts, T.H., Küper, Kreft, H., Barthlott, W. (2005): Global patterns of plant diversity and floristic knowledge. – Journal of Biogeography 32(7): 1107-1116.

[5] Dinerstein, E., Vynne, C., Sala, E., Joshi, A.R., Fernando, S., Lovejoy, T.E., Mayorga, J., Olson, D., Asner, G.P., Baillie, J.E.M., Burgess, N.D., Burkart, K., Noss, R.F., Baccini, A., Birch, T., Hahn, N., Joppa, L.N., Wikramanayake, E. (2019): A global Deal for nature: Guiding principles, milestones, and targets. – Science Advances 5(4): eaaw2869.

[6] CONABIO. (1998): La diversidad biológica en México: Estudio de País, 1998. – Comisión Nacional para el Conocimiento y Uso de la Biodiversidad. México.

[7] Villarreal-Quintanilla, J.A., Bartolomé-Hernández, J.A., Estrada-Castillón, E., Ramírez-Rodríguez, H., Martínez-Amador, S. J. (2017): The endemic element of the Chihuahuan Desert vascular flora. – Acta Botanica Mexicana 118: 65-96.

[8] Riemann, H., Exequiel, E. (2007): Endemic regions of the vascular Flora of the peninsula of Baja California, Mexico. – Journal of Vegetation Science 18(3): 327-336.

[9] Villaseñor, J.L., Ortiz, E., Delgadillo-Moya, C., Juárez, D. (2020): The breadth of the Mexican transition zone as defined by its flowering plant generic flora. – PLoS ONE 15(6): e0235267.

[10] López-González, C., Presley, S.J., Lozano, A., Stevens, R.D., Higgins, C.L. (2015): Ecological biogeography of Mexican bats: The relative contributions of habitat heterogeneity, beta diversity, and environmental gradients to species richness and composition patterns. – Ecography 38: 261-272.

[11] Fernández, J.A. (2012): Phylogenetics and biogeography of the microendemic rodent Xerospermophilus perotensis (Perote ground squirrel) in the oriental basin of Mexico. – Journal of Mammalogy 93(6): 1431-1439.

[12] Fernández, J.A., Cervantes, F.A., Hafner, M.S. (2012): Molecular systematics and biogeography of the Mexican endemic kangaroo rat, Dipodomys phillipsii (Rodentia: Heteromyidae). – Journal of Mammalogy 93(2): 560-571.

[13] Moonlight, P.W., Richardson, J.E., Tebbitt, M.C., Thomas, D.C., Hollands, R., Peng, C.I., Hughes, M. (2015): Continental-scale diversification patterns in a megadiverse genus: The biogeography of neotropical begonia. – Journal of Biogeography 42: 1137-1149.

[14] Chazdon, R.L., Harvey, C.A., Martínez-Ramos, M., Balvanera, P., Schondube, J.E., Stoner, K.E., Cabadilla, L.D.A., Flores-Hidalgo, M. (2011): Seasonally dry tropical Forest biodiversity and conservation value in agricultural landscapes of Mesoamerica. – In: Dirzo,

R., Young, H.S., Mooney, H.A., Ceballos, G. (eds.) Seasonally Dry Tropical Forests. Island Press, Washington, DC.

[15] Estado de la Región. (2008): Estado de la Región en Desarrollo Humano Sostenible un informe desde Centroamérica y para Centroamérica. – Programa Estado de la Región. San José, Costa Rica.

[16] Bastida-Zavala, J.R., del Socorro García-Madrigal, M., Rosas-Alquicira, E.F., López-Pérez, R.A., Benítez-Villalobos, F., Meraz-Hernando, J.F., Torres-Huerta, A.M., Montoya-Márquez, A., Barrientos-Luján, N.A. (2013): Marine and coastal biodiversity of Oaxaca, Mexico. – Check List 9(2): 329-390.

[17] de León, G.P.P., García-Prieto, L., Mendoza-Garfia, B. (2011): Describing parasite biodiversity: The case of the helminth Fauna of wildlife vertebrates in Mexico. – In: Grillo, O., Venora, G. Changing Diversity in Changing Environment. BoD – Books on Demand.

[18] Hershler, R., Liu, H.P., Landye, J.J. (2011): New species and records of springsnails (Caenogastropoda: Cochliopidae: Tryonia) from the Chihuahuan Desert (Mexico and United States), an imperiled biodiversity hotspot. – Zootaxa 3001: 1-32.

[19] Sáenz-Romero, C., Rehfeldt, G.E., Crookston, N.L., Duval, P., St-Amant, R., Beaulieu, J., Richardson, B.A. (2010): Spline models of contemporary, 2030, 2060 and 2090 climates for Mexico and their use in understanding climate-change impacts on the vegetation. – Climatic Change 102: 595-623.

[20] Silva-Flores, R., Pérez-Verdín, G., Wehenkel, C. (2014): Patterns of tree species diversity in relation to climatic factors on the Sierra Madre occidental, Mexico. – PLOS One 9(8): e105034.

[21] Mosco, A. (2017): Niche characteristics and potential distribution of Thelocactus species, a Mexican genus of globular cacti. – doi:10.1101/124511. PPR:PPR32150.

[22] Peters, R., Ripple, W.J., Wolf, C., Moskwik, M., Carreón-Arroyo, G., Ceballos, G., Córdova, A., Dirzo, R., Ehrlich, P.R., Flesch, A.D., List, R., Lovejoy, T.E., Noss, R.F., Pacheco, J., Sarukhán, J.K., Soulé, M.E., Wilson, E.O., Miller, J.R.B., and 2556 scientist signatories from 43 countries. (2018): Nature Divided, Scientists United: US–Mexico Border Wall Threatens Biodiversity and Binational Conservation. – BioScience 68(10): 740-743.

[23] Lasky, J.R., Jetz, W., Keitt, T.H. (2011): Conservation biogeography of the US–Mexico border: A transcontinental risk assessment of barriers to animal dispersal. – Diversity and Distributions 17: 673-687.

[24] Dávila, P., Benítez, H., Barrios, Y., Cruz-Angón, A., Álvarez-Girard, N. (2011): Definition and insertion of the GSPC in the political context of Mexico. – Botanical Journal of the Linnean Society 166(3): 326-330.

[25] Fodor, R.K. (2014): Ökoszisztéma szolgáltatások egy újfajta keretrendszerben. – In: Lukovics, M., Zuti, B. A területi fejlődés dilemmái. SZTE Gazdaságtudományi Kar, Szeged.

[26] Alix-Garcia, J.M., Sims, K.R.E., Yañez-Pagans, P. (2015): Only one tree from each seed? Environmental effectiveness and poverty alleviation in Mexico's payments for ecosystem. – American Economic Journal: Economic Policy 7(4): 1-40.

[27] Costedoat, S., Corbera, E., Ezzine-de-Blas, D., Honey-Rosés, J., Baylis, Castillo-Santiago, M.A. (2015): How Effective Are Biodiversity Conservation

Payments in Mexico? – PLoS ONE 10(3): e0119881.

[28] Honey-Rosés, J., Baylis, K., Ramírez, M.I. (2011): A spatially explicit estimate of avoided forest loss. – Conservation Biology 25(5): 1032-1043.

[29] Durand, L., Vázquez, L.B. (2011): Biodiversity conservation discourses. A case study on scientists and government authorities in sierra de Huautla biosphere reserve, Mexico. – Land Use Policy 28: 76-82.

[30] Mendez-Lopez, M.E., García-Frapolli, E., Pritchard, D.J., Gonzalez, M.C.S., Ruiz-Mallen, I., Porter-Bolland, L., Reyes-Garcia, V. (2014): Local participation in biodiversity conservation initiatives: A comparative analysis of different models in south East Mexico. – Journal of Environmental Management 145: 321-329.

[31] Torres-Miranda, A., Luna-Vega, I., Oyama, K. (2011): Conservation biogeography of red oaks (Quercus, section Lobatae) in Mexico and Central America. – American Journal of Botany 98(2): 290-305.

[32] Sanborn, A.F., Phillips, P.K. (2013): Biogeography of the cicadas (Hemiptera: Cicadidae) of North America, North of Mexico. – Diversity 5: 166-239.

[33] Moreno-Calles, A., Casas, A., Blancas, J., Torres, I., Masera, O., Caballero, J., Garcia-Barrios, L., Perez-Negron, E., Rangel-Landa, S. (2010): Agroforestry systems and biodiversity conservation in arid zones: The case of the Tehuacan Valley, Central Mexico. – Agroforestry Systems 80: 315-331.

[34] Vallejo, M., Casas, A., Blancas, J., Moreno-Calles, A.I., Solís, L., Rangel-Landa, S., Dávila, P., Téllez, O. (2014): Agroforestry systems in the highlands of the Tehuacán Valley, Mexico: Indigenous cultures and biodiversity conservation. – Agroforestry Systems 88: 125-140.

[35] Robson, J.P., Berkes, F. (2011): Exploring some of the myths of land use change: Can rural to urban migration drive declines in biodiversity? – Global Environmental Change 21: 844-854.

[36] Larios, C., Casas, A., Vallejo, M., Moreno-Calles, A.I., Blancas, J. (2013): Plant management and biodiversity conservation in Náhuatl homegardens of the Tehuacán Valley, Mexico. – Journal of Ethnobiology and Ethnomedicine 9:74.

[37] Ruben, R., Zuniga, G. (2010): How standards compete: Comparative impact of coffee certification schemes in northern Nicaragua. – Supply Chain Management: An International Journal 16(2): 98-109.

[38] Tejeda-Cruz, C., Silva-Rivera, E., Barton, J.R., Sutherland, W.J. (2010): Why shade coffee does not guarantee biodiversity conservation. – Ecology and Society 15(1): 13.

[39] Solís-Marín, F.A., Honey-Escandón, M.B.I., Herrero-Perezrul, M.D., Benitez-Villalobos, F., Díaz-Martínez, J.P., Buitrón-Sánchez, B.E., Palleiro-Nayar, J.S., Durán-González, A. (2013): The echinoderms of Mexico: Biodiversity, distribution and current state of knowledge. – In: Alvarado, J.J., Solís-Marín, F.A. (eds.) Echinoderm Research and Diversity in Latin America. – Springer-Verlag, Berlin.

[40] Wong-Arguelles, C., Alonso-Castro, A.J., Ilizaliturri-Hernandez, C.A., Carranza-Alvarez, C. (2020): Credibility of In Situ Phytoremediation for Restoration of Disturbed Environments. – In: Hakeem, K., Bhat, R., Qadri, H. Bioremediation and Biotechnology. Springer, Cham.

[41] del Pilar Ortega-Larrocea, M., Xoconostle-Cazares, B.,

Maldonado-Mendoza, I.E., Carrillo-Gonzalez, R., Hernandez-Hernandez, J., Garduño, M.D., Lopez-Meyer, M., Gomez-Flores, L., del Carmen A. Gonzalez-Chavez, M. (2010): Plant and fungal biodiversity from metal mine wastes under remediation at Zimapan, Hidalgo, Mexico. – Environmental Pollution 158: 1922-1931.

[42] Coleman-Derr, D., Desgarennes, D., Fonseca-Garcia, C., Gross, S., Clingenpeel, S., Woyke, T., North, G., Visel, A., Partida-Martinez, L.P., Tringe, S.G. (2016): Plant compartment and biogeography affect microbiome composition in cultivated and native agave species. – New Phytologist 209: 798-811.

[43] Panke-Buisse, K., Poole, A.C., Goodrich, J.K., Ley, R.E., Kao-Kniffin, J. (2015): Selection on soil microbiomes reveals reproducible impacts on plant function. – ISME Journal 9: 980-989.

[44] Turner, T.R., James, E.K., Poole, P.S. (2013): The plant microbiome. – Genome Biology 14: 209.

[45] Folaranmi, J. (2013): Production of Biodiesel (B100) from Jatropha Oil Using Sodium Hydroxide as Catalyst. – Journal of Petroleum Engineering Article ID: 956479.

[46] Fresnedo-Ramírez, J., Orozco-Ramírez, Q. (2013): Diversity and distribution of genus Jatropha in Mexico. – Genetic Resources and Crop Evolution 60: 1087-1104.

Chapter 2

Ecology of Plant Communities in Central Mexico

Joaquín Sosa-Ramírez, Vicente Díaz-Núñez and Diego R. Pérez-Salicrup

Abstract

In Central Mexico converge three biogeographic provinces: Altiplano sur, Sierra Madre Occidental and Costa del Pacífico. Each one of them is composed by different plant communities: Thorn Forest, Temperate Mountain Forest and Dry Tropical Forest respectively. Our objective is to show, through phytoecological analysis, the species richness, diversity and the structure of the plant communities from the Temperate Mountain Forest and from the Tropical Dry Forest. In the Temperate Mountain Forest, 50 forest species were recorded, with a Shannon Wiener diversity index H′ = 1.63 on altitudes from 2400 to 2600 m. The Whittaker β index is B_w = 7.22. In the tropical dry forest, we identified 79 plants species with a mean diversity index H′ = 3.49 on altitudes from 1951 to 2100 m. In this ecosystem the B_w index is 8.12. This study offers important information for the establishment of management practices, considering the protection status from the areas in which this vegetation type is distributed.

Keywords: Aguascalientes, Sierra Fria, Temperate Mountain Forest, Tropical Dry Forest, biogeographic provinces

1. Introduction

Mexico is one of the five countries with the greatest biological diversity in the world, due, in part to the confluence of the Neartic (North America) and Neotropical biogeographic zones (Mexico, Central and south America). As well as, the species evolutionary processes in its territory [1]. The Mexican territory represent only 1% of the earth's surface; nevertheless, Mexico belongs to the select group of the five countries considered megadiverse, along with Brazil, Colombia, China and Indonesia [2, 3]. Due to its geographic locations and its multiple landscapes, a large number and diversity of ecosystems converge in the national territory. For that reason, Mexico is ranked 12th in terms of global forest area [4]. Even though, multiple efforts have been made for the forest conservation during the last decade of the XXI century, on a global scale, forest have been transformed to other uses at a rate of $1.3x10^6$ million ha/yr. or they have been affected by natural disturbances that have partially or totally changed their structure. This amount represents a 19% decrease in comparison to the exchange rate registered in the last decade of the 20th century ($1.6X10^6$ million ha/yr) [5]. Temperate forests in Mexico are found mostly, although not exclusively, in the mountainous areas along the Sierra Madre Occidental (the area with the highest concentration of forest ecosystems in the country), the mountains of Sierra Madre Oriental, the Sierra Norte de Oaxaca and the Altos de Chiapas, as well as in different mountain ranges and isolated mountains in the Altiplano and intermingled in the tropical plains [6]. The conifer and oak forest in Mexico represent the most extensive vegetation cover in terms of vegetation types dominated by woody species, this species covers 16.4% of the total surface of the country, being only surpassed by the xeric shrubland which is the vegetation type which has the largest extension [5]. These ecosystems are important both economically and ecologically, since they support productive activities, harbor great biological diversity and serves as a refuge for wildlife. Likewise, forest provide essential environmental goods and services for the human society subsistence [7, 8].

The State of Aguascalientes has a total extension of 555, 867.4 hectares, of which 291,792.4 hectares equivalent to 52.5% present some forest type [9]. According to the classifications issued by different sources [10, 11], the State of Aguascalientes is made up by three large ecoregions (biogeographic regions), the Temperate Mountain Forest, the Tropical Dry Forest (also known as lowland deciduous forest) and the thorn forest (including crasicaule shrubland and xeric shrubland). The first ecosystem type is mainly distributed in la Sierra Fria, Sierra del Laurel, Sierra de Tepezalá and Cerro de Juan el Grande in El Llano municipality (**Figure 1**). The largest area covered by Temperate Mountain Forest vegetation in Aguascalientes is located in an area locally known as Sierra Fria, this site is a Protected Natural Area by state and federal decree which covers close to 107,000 ha [12]. In the Temperate Mountain Forest, the plant communities the most common vegetation types are oak forests (*Quercus* spp.), pine trees (*Pinus* spp.), oak-pine, pine-oak, juniper (*Juniperus* spp.), manzanita shrubland (*Arcostaphyllos pungens*) and different associations of these genera. The vegetation that has mainly colonized the sites that had been disturbed are *Juniperus deppeana* and *A. pungens*, although there has also been an increase in conifer populations [13]. The second largest formation where this ecosystem is found is located in la Sierra del Laurel in the Southwest corner of the State occupying close to 17,000 ha. This area presents similar plant communities

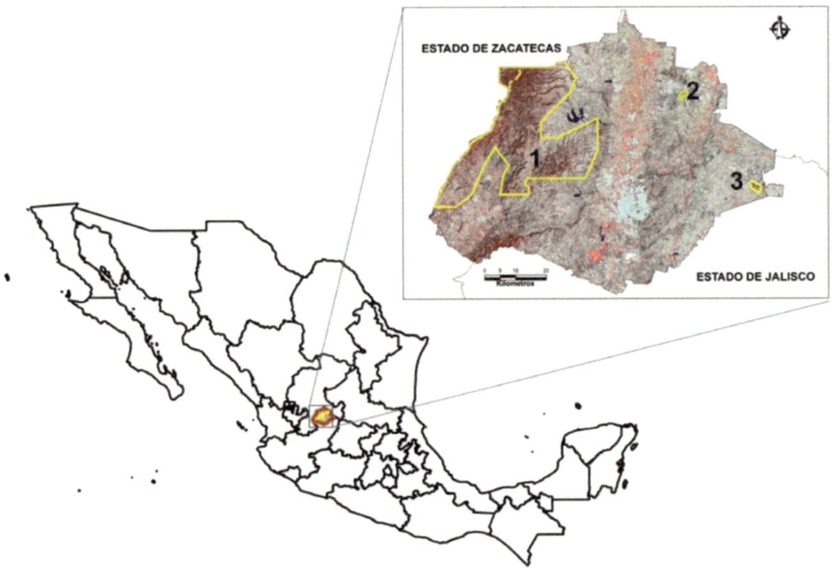

Figure 1.
Distribution areas of the Temperate Mountain Forest in Aguascalientes state. (1) sierra Fria protected natural area (San José de Gracia, Pabellón de Arteaga, Rincón de Romos and Calvillo municipalities); (2) Mountain Hill of Tepezalá, and (3) Juan el Grande Mountain (El llano municipality).

but with a greater dominance of oak populations (*Quercus* spp.) and lower density of manzanita (*A. pungens*).

The largest area occupied by the tropical dry forest is mainly located in the Calvillo municipality, although, there are relics of vegetation indicative of this ecosystem in the Jesus Maria, San Jose de Gracia and Aguascalientes municipalities, which suggests a larger presence of this vegetation type in the past. In the tropical dry forest, forest structures made up of shrubs and trees between 2 and 8 m high and some relics of medium tropical forest. In Aguascalientes, this is one of the ecosystems with the highest species richness [14]. The most representative vegetation in this ecoregion corresponds mainly to the *Lysiloma*, *Bursera*, *Ipomoea*, *Acacia*, *Eysenhardthia*, *Opuntia*, *Mimosa* and *Agave* genera.

Our objective was to provide an overview of some ecological aspects (species richness, diversity and distribution) of woody species natural communities in the most representative ecosystems of the State of Aguascalientes, assuming that there would be a high similarity degree with the vegetation of neighboring sites, considering both the environmental and physiographic characteristics from this State.

2. Materials and methods

Three studies were conducted individually. During 2008–2015, the natural communities of the temperate mountain forest in the area commonly known as Sierra Fria, in the northwest of the State of Aguascalientes, as well as the main disturbances that have affected them in the past and present were analyzed [10, 13]. Likewise, during the period 2011–2015 a study was carried out to determine the diversity, dynamics and functioning of the tropical dry forest in the Calvillo municipality [14, 15].

2.1 Temperate Mountain Forests

2.1.1 Study area and sampling design

This study was carried out in to the Sierra Fria Protected Natural Area (SF-PNA) which is 106,114.6 hectares in size and is located in the northwest of the Aguascalientes State. This area has an altitude ranging between 2,100 and 3,050 masl. The study area comprised 25 thousand hectares, in a polygon located between the coordinates 102°31′31″ to 102°37′44″ west longitude and 22°05′47″ at 22°14′03″ north latitude, assuming that the conditions both geographic, ecological and climatic are representative of the entire ANP (See **Figure 2**).

A stratified sampling strategy was developed [16]. The sampling strata were delimited based on the altitude, solar exposure, and geoform of the site (flat, concave and convex terrain). The first stratum was defined using a Digital Elevation Model (DEM) of the ANP SF, elaborating a spatial grid according to five altitudinal categories: i) 2,000-2,200, ii) 2,200-2,400, iii) 2,400-2,600, iv) 2,600-2,800, and v) >2 800 masl.

To stablish the altitudinal strata, the level curves from study site were defined using the DEM. The solar exposure was approached using an exposure map made with a SPOT 2010® imagine on which the DEM of the site was superimposed. Subsequently, a mesh map was prepared using the ArcGis 10.2. The geoform was obtained based on the slope, where flat terrain = sites with a slope $\leq$ 10%, concave t. = slope $\geq$ 10 and $\leq$ 25% and convex t. = slope $\geq$ 25%.

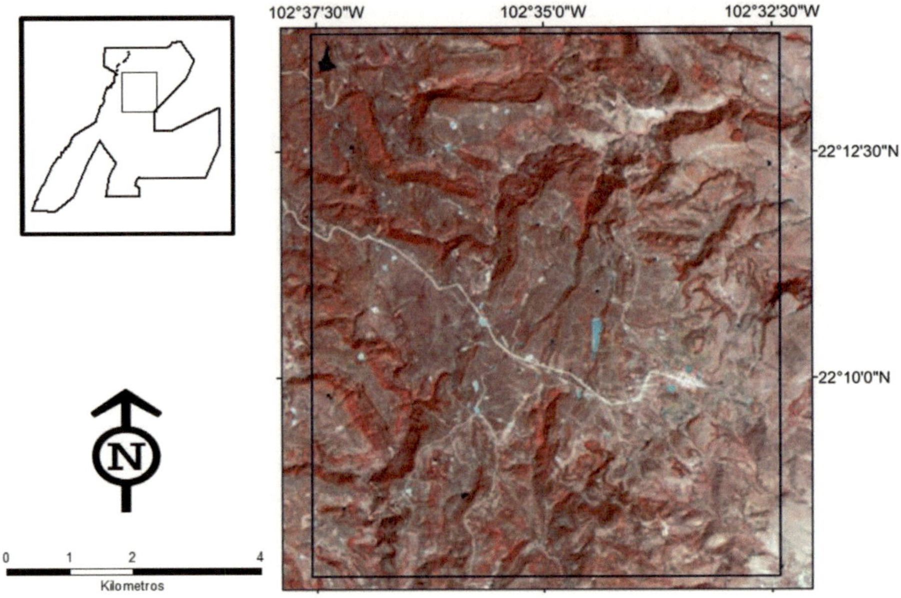

Figure 2.
Location of the protected natural area sierra Fria, the study area of the Temperate Mountain Forest.

2.1.2 Identification, distribution and abundance of forest species

To identify the tree and shrub diversity in the study area, we conducted 60 phytoecological inventories in 60 different sites distributed randomly using the sampling scheme already described (**Table 1**).

The field samplings were performed in rectangular plots of 600 m^2, with a central line 100 m in length and two lateral lines with three m of separation. In each inventory, the frequency of the tree and shrub species present were determined, as well as the site environmental variables. Individuals with DBH $\geq$ 5 cm and height $\geq$ 1.50 m were considered as trees. Individuals below these categories were

Altitude levels	**Topographic position**									
	Concave				**Convexe**				**Flat**	**Total**
	N	**S**	**E**	**W**	**N**	**S**	**E**	**W**		
2000–2200	0	0	0	0	1	1	0	0	1	3
2200–2400	0	1	0	0	3	2	0	0	0	6
2400–2600	8	4	1	1	8	3	0	0	3	28
2600–2800	2	0	0	1	4	3	0	0	11	21
> 2800	2	0	0	0	0	0	0	0	0	2
Total	**12**	**5**	**1**	**2**	**16**	**9**	**0**	**0**	**15**	**60**
Total inventories					**60**					

The intersections between lines and columns whose value is zero, indicate areas with little representativeness in the landscape and consequently an absence of samplings.

Table 1.
Number of samplings performed at different altitudinal levels, topographic positions and solar exposures, derived from the sampling system.

considered as juveniles and shrubs. The variables recorded in the site were: altitude, slope (in %), solar exposure (N, S, E, W), physiography (flat land, hillock, plateau, middle slope, high slope, ravine bottom, creek), coverage (c1 = ≤10%; c2 = 11–30%; c3 = 31–50%; c4 = 51–70% y c5 = ≥70%) and geoform. Management variables related to land use (no use, forest exploitation, wildlife management, grazing, agriculture and conservation) were considered as well as intensity of use (null, moderate, over-exploited and not determinable). Each one of the sampling points were geographically located by Transverse Mercator Units (UTM).

In order to identify the oak and conifer species in the field, keys generated by De la Cerda [17] and Siqueiros [18], respectively, were used. The unknown species were collected in botanical presses and identified at the Autonomous University of Aguascalientes herbarium (HUAA). To leave evidence of the new species records in the ANP SF, specimens were deposited in the HUAA.

2.1.3 Distribution and abundance of species

To estimate the distribution of tree and shrub forest species, the presence of each of the species found in each of the 60 sampling sites was quantified. In the case of species considered as restricted distribution (eg. *Quercus cocolobifolia, Pinus chihuahuana*, and *P. duranguensis* var. *quinquefoliata*), samples were taken at specific sites (n = 4), according to the information provided by De la Cerda [17] and Siquéiros [18]. Species with a wide distribution were those that occurred in the greatest number of sites.

The frequency of the species found was determined on 100 m transect at ground level, observing 100 separate points every meter. The species found at each point were recorded (when there was more than one vegetation layer), counting the number of times that each species appeared (absolute frequency) [16] over the whole transect. Relative frequency was calculated using the Equation [19]:

$$\text{Relative frequency} = \left(\frac{\text{Species frequency}}{\sum \text{Frequency values of all species}}\right) \tag{1}$$

Where:

Frequency of the species x = absolute frequency obtained from each site sampling.

Subsequently, an abundance index was calculated using the equation:

$$\text{Spp.ai} = \frac{\sum_{\text{relative frequencies}}}{\text{Number of sampled sites}} \tag{2}$$

Where:

Spp.ai = Identified Species abundance index.

With this data, distribution and abundance graphs of the main arboreal-shrub forest species were created. The phytoecological analysis was used to calculate the species richness and the Shannon index diversity (H') and the beta Whittaker's (ßw) index respectively, the first were calculated as a function of the altitudinal level, the second also incorporating the geoform using the Species Diversity and Richness® (Pisces Conservation LTD) software. Pear calculate the indexes we used the equation:

$$H' = -\sum_{i=1}^{S} p_i \log_2 p_i \tag{3}$$

Where

S = species richness; Pi = proportion of the individuals of species i with respect to the total number of individuals; ni = number of individuals of species i

$$ß_w = \frac{S}{\overline{S}} \quad (4)$$

Where:

S = Species richness and S = mean richness of the site.

2.2 Dry Tropical Forest (DTF)

2.2.1 Study area

Although there are some studies that suggest the existence of relics of Dry Tropical Forest (DTF) vegetation in some municipalities of the Aguascalientes State [15, 20], this ecosystem has a greater representation both in surface area and in its conservation status in Calvillo municipality. The study was conducted in 26 sites with DTF vegetation cover in Terrero de la Labor ejido, located within the Sierra Fria Protected Natural Area, in the Municipality of Calvillo, State of Aguascalientes, in Central Mexico. The ejido polygon comprises an area of 5,861 ha. [21], of which, the DTF occupies 45% of its total area (**Figure 3**). It is located within the extreme coordinates: 102°43′58.88“ West Longitude and 22°6’4.78” North Latitude and at the Southeast end 102°41′24.95“ West Longitude and 21°44’27.61” North Latitude.

2.2.2 Selection of the study sites and sampling design

We used a stratified sampling design system [16]. Sampling strata were delimited based on geoforms, slope, exposure and altitude. To characterize geoforms, three criteria were used: concave, convex and flat terrain. A concave geoform was defined when the slope ranged between 10 and 25%, which usually corresponded to ravines and small depressions. When the sites had a slope between 25 and 60% they were characterized as convex sites. Flat terrains had slopes <10%.

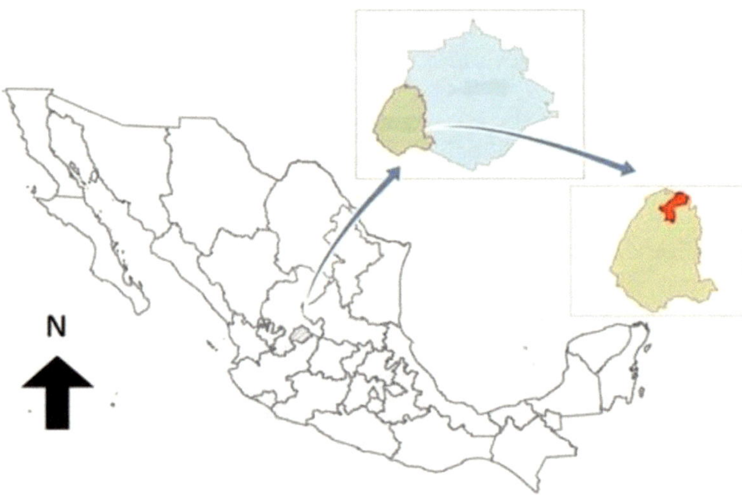

Figure 3.
Location of the study area. (A) Mexico, (B) state of Aguascalientes, (C) municipality of Calvillo and (D) Terrero de la labor Ejido.

Solar exposure was defined using an exposure map made with a Geographic Information System from a 2008 Spot® satellite image and a digital elevation model (MDE). Only the main cardinal points (North, South, East and West) were considered. To locate the altitudinal strata, the contours of the zone defined from the MDE were used. Subsequently, a grid map was developed for the identification of the sampling areas (See **Figure 4**).

2.2.3 Selection and characterization of sites to quantify of the composition and abundance of woody species

We established 26 sites to quantify phytoecological inventories, distributed in the landscape according to the above mentioned sampling system. At each point, the projected coordinates of the site were taken with GPS Garmin 48 XL line in UTM format, zone 13 North and with reference Datum WGS84 and with accuracies of 5 to 12 m with differential kinematic adjustment (WAAS). Subsequently, the points were placed on a SPOT 2010 satellite image (**Figure 5**). Site variables considered were the slope (%), solar exposure, physiography of the terrain, intensity and type of use and canopy coverage.

Slope at each sampling site was obtained by direct field measurement with a Bruntton clinometer with a precision of +/− 2° of variation for each 100 meters of length. This data in turn was contrasted with the data obtained from the digital elevation model with precision of 1 to 2 meters in the Z value. Five classes were used to define the slope: i) <10%, ii) 11–30, iii) 31–50, iv) 51–70 and v) > 70. Exposure to solar radiation was estimated considering the cardinal points North (N), South (S), East (E) and West (O), as well as their combinations.

The altitude of each site was obtained directly in the field with the support of a GPS with barometric adjustment to reduce the effect of mathematical variation of the Geoid model and with precision of 1 to 3 meters. This was compared with the data obtained from the prospecting of points against elevation level curves obtained

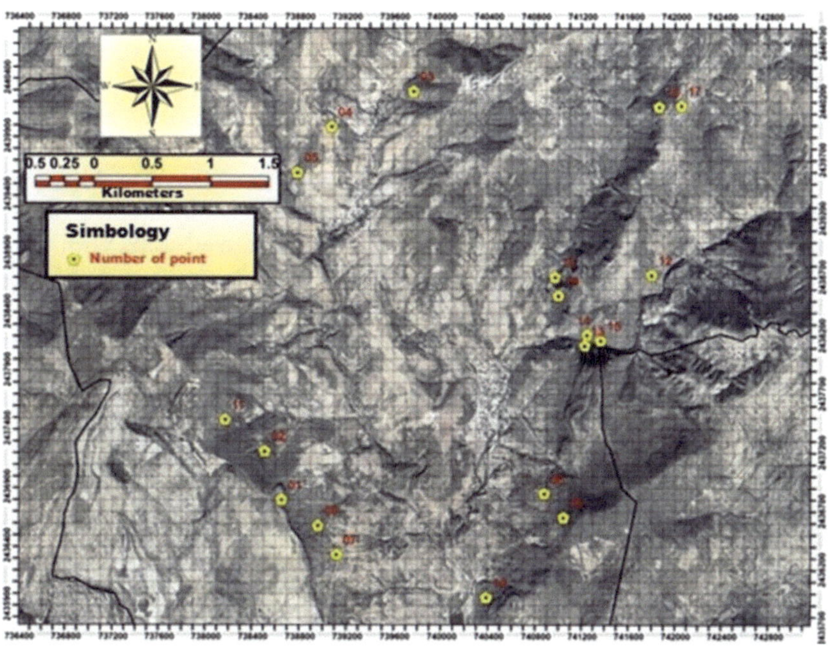

Figure 4.
Geographic representation of dry tropical Forest and the sampling points in Terrero de la labor Ejido, in the municipality of Calvillo, Aguascalientes.

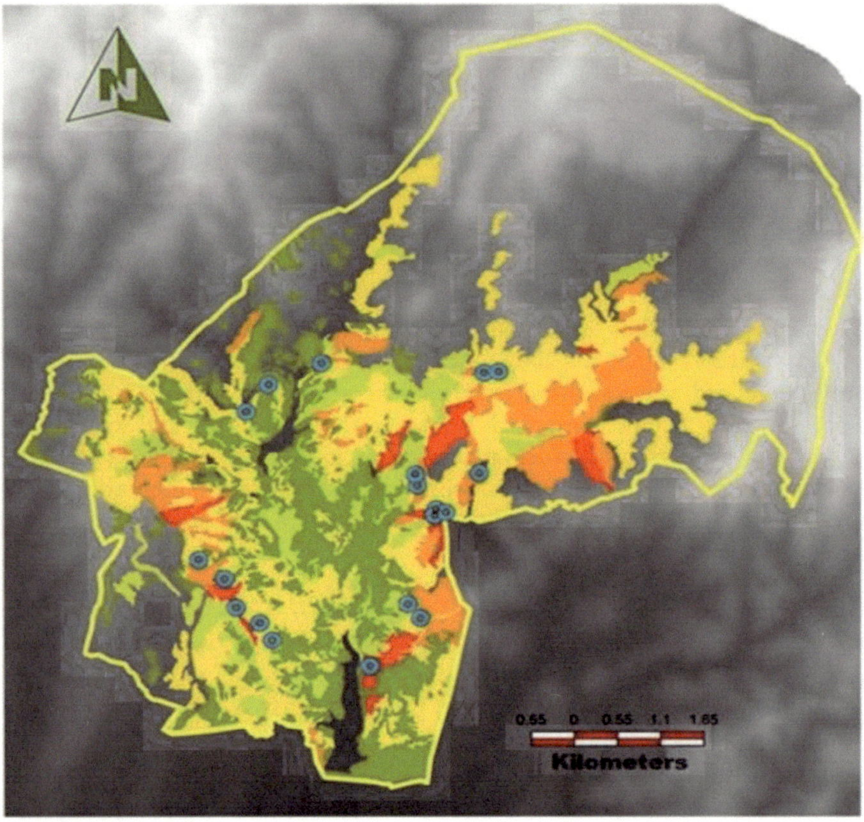

Figure 5.
Ipsographic model of Ejido Terrero de la labor Ejido polygon, and distribution of the sampling points in the DTF.

from the digital elevation model to reduce the potential errors of direct measurements.

The physiography of the terrain was characterized considering flat terrain (slope < 10%), steep (without slope), medium slope (10–25%) and high slope (>60%). The exposure of the sites was quantified with a compass and the magnetic north was taken as reference for its definition in the previously defined ranges. Exposure for each stand of the sampling site was also analyzed along with the digital model of exposures generated from the digital elevation model. The **Table 2** shown the sample points distributed in the landscape of the Dry Tropical Forest.

Other characteristics considered in the description of the sites were the degree of modification (i.e. transformation of geographical space, introduction of species), its intensity (light, medium and overexploited), as well as the type of use by local inhabitants (hunting, grazing, gathering, etc.).

2.2.4 Species richness

To describe species composition, we used a sampling design based on nested plots in an area of 1024 m^2 in each inventory, using the criteria of the minimum area [16]. We started with a plot of 1 x 1 m in a direction perpendicular to the slope in which all present species were recorded, and subsequently, the plot. Subsequently, the plot was increased in size to 2 X 1, 2 X 2, 2 X 4, 4 X 4 m etc. registering the new species for each increment in the area of the squares until reaching the maximum

Meters above sea level (masl)	Topographic position										Total
	Concave coverage (%)					Convex coverage (%)					
	<10	11–30	31–50	51–70	>71	<10	11–30	31–50	51–70	>71	
1800–1900	0	0	0	0	1	0	1	0	3	0	5
1901–2000	0	0	0	0	0	1	1	3	5	1	11
2001–2100	0	0	0	0	0	1	1	2	2	0	6
>2100	0	0	0	0	0	1	1	0	4	0	4
Total inventories	0	0	0	0	1	0	0	5	14	1	26

Table 2.
Distribution of samplings sites according to the proposed design.

extension (i.e.: 32 x 32 m = 1024 m2), to obtain an area/species curve. We then identified the area in which the present species stabilized. This sampling method increased the probability of finding rare species as the area increased, an effect known as Rarefaction [22].

Identification of species was estimated in the field by morphological characters described in previous studies. Specimens that could not be identified in the field were collected and later identified in the Herbarium of the Autonomous University of Aguascalientes (HUAA).

We used the linear intercept survey method (Canfield line). A 100 m long line was perpendicular to the slope, starting at the GPS coordinates of the sampling site, then intersection lines were defined were individuals of DTF species were counted at constant intervals of 1 meter. Shrub and tree individuals were categorized into five heights classes 0–1 m, 1.1–2 m, 2.1–4 m, 4–8 m, 8–15 m and > 16 m. For each class we measured canopy cover of each species by measuring the perpendicular projection of the crown and the frequency of species. To estimate crown, cover the following formula was used:

$$\text{Cover (C)} = \Sigma \text{ length of individuals of species i/total length of intersections X 100.}$$

To estimate frequencies, we used the formula:

$$\text{Frequency (F)} = \Sigma \text{ of number of times that individuals of the species intercepted by the line}/\Sigma \text{ total species intercepted X 100.}$$

2.2.5 Data analysis

Species composition was estimated through the identification of the species found in each of the sampling plots. To find a limit on the number of samples and to reduce the possibility of under- or over-sampling, we conducted a rarefaction analysis. The Shannon-Wienner alpha diversity (H') was calculated for each of the sites and for each altitudinal level using the Richness and diversity species® software, considering that there could be variation in diversity according to the change in environmental conditions in temperature and precipitation as mentioned in the Standard Atmospheric Index (decrease of 0.6°C/100 m altitude).

The formula of the Shannon index is:

$$H' = -\sum_{i=1}^{S} p_i \log_2 p_i \tag{5}$$

where:

- S– Total number of species (species richness)
- p_i– Proportion of individuals of species i in respect to total of individuals (i.e.: relative abundance of species i): $\frac{n_i}{N}$
- n_i– number of individuals of species i
- N– Total number of individuals of all species

The index considers the number of species present in the study area (species richness), and the relative number of individuals of each of those species (abundance).

To estimate replacement rates of species Whitakker's β diversity was computed, using the diversity found for each altitudinal level analyzed as reference.

$$\beta = \frac{S}{\alpha - 1} \tag{6}$$

Where: β = Whitakker's β diversity.
S = Total number of species in samples.
α = Mean number of species in samples.

3. Results

3.1 Temperate Mountain Forest

3.1.1 Richness and diversity species

In the 60 sites, 50 species were recorded, corresponding to 20 families and 27 genera (**Table 3**), of which, due to their structure, 47% (n = 24) were considered trees (height ≥ 3.5 m) and 53% (n = 27) shrubs and juveniles. The best represented families were *Fagaceae* (11 species), *Pinaceae* (8 species) and *Ericaceae* (5 species). The species *Q. obtusata* Bonpl., *J. duranguensis* Martínez and *Crataegus* sp. are new reports in the Sierra Fria.

On average, the highest H' diversity index is found in sites whose altitude ranges between 2400 and 2600 and 2600–2800 mamsl (H' = 1.48 y 1.63, respectively), the former associated with ravines and difficult access places; the second index corresponds to places with higher moisture content and without use. The lowest indexes (H = 1.22 y 1.36) were found in altitudinal ranges of 2200–2400 and 2000–2200 m, respectively, located on flat lands with intensive management and high resource use rates. According to the geoform, the diversity Wittaker's *ß* was greater on the convex sites (*ßw* = 5.80), followed by the concave sites (*ßw* = 4.27) and flat lands (*ßw* = 4.04). According to the altitudinal level, the highest diversity was found in the sites whose altitude ranges between 2,400 and 2,600 m (*ßw* = 7.22), mainly in ravines and places hard to access. In contrast, the lowest indexes were found on site with an altitude lower than 2, 400 m (*ßw* = 4.52), located on flat lands, under intensive management and easy access.

In the **Figure 6**, we shown an example of dominant vegetation in Temperate Mountain Forest (in conifers, *Pinus leiophylla* and *P. teocote* in order) in the Sierra Fria Protected Natural Area.

Species	Key	Common name*	Family	Forest classification**	Use¥	Report***
Acacia farnesiana	*Acafar*	Huizache	*Leguminosae*	Tr	Nu	Y
Asclepias linearis	*Aline*	Romerillo	*Apocynaceae*	Sh	Nu	Y
Arbutus arizonica	*Aariz*	Madroño	*Ericaceae*	Sh	Fe	Y
Arbutus xalapensis	*Axala*	Madroño rojo	*Ericaceae*	Sh	Fe	Y
Arbutus glandulosa	*Aglan*	Madroño blanco	*Ericaceae*	Sh	Fe	Y
Arctostaphylos pungens	*Apun*	Manzanita	*Ericaceae*	Sh	Fe	Y
Budleia scordioides	*Bsco*	Vara blanca	*Compositae*	Sh	Fr	Y
Budleia cordata	*Bcor*	Tepozan	*Compositae*	Tr	Fr	Y
Bursera fagaroides	*Burfaga*	Venadilla	*Burseraceae*	Sh	Nu	Y
Comerostaphyllis spp.	*Comesp*	Pacuato	*Ericaceae*	Sh	Med	Y
Dalea bicolor	*Dabic*	Engordacabra	*Fabaceae*	Sh	Fr	Y
Dasylirion acotriche	*Dasaco*	Sotol	*Agavaceae*	Sh	Nu	Y
Dodonaea viscosa	*Dovisc*	Jarilla	*Sapindaceae*	Sh	Med	Y
Eucaliptus camaldulensis	*Eucamal*	Eucalipto	*Myrtaceae*	Tr	Nu	Y
Fraxinus uhdei	*Frauhd*	Fresno	*Oleaceae*	Tr	Nu	Y
Garria ovata	*Garova*	planta peluda	*Garryaceae*	Sh	Nu	Y
Pinus chihuahuana	*Pinchi*	Pino Prieto	*Pinaceae*	Tr	Nu	Y
Pinus duranguensis Mart.	*PiduM*	Pino verde	*Pinaceae*	Tr	Ew	Y
Pinus duranguensis f. quinquefoliata	*PiduQ*	Pino verde	*Pinaceae*	Tr	Nu	Y
Pinus leiophylla	*Pile*	Pino Prieto	*Pinaceae*	Tr	Nu	Y
Pinus lumholtzii	*Pilum*	Pino llorón	*Pinaceae*	Tr	Nu	Y
Pinus michoacana	*Pimich*	Pino barbón	*Pinaceae*	Tr	Nu	Y
Pinus cembroides	*Picem*	Pino chaparro	*Pinaceae*	Tr	Nu	Y
Prosopis laevigata	*Prolae*	Mesquite	*Pinaceae*	Sh	Nu	Y
Pinus teocote	*Pinteo*	Pino	*Pinaceae*	Tr	Ew	Y
Jatropha dioica	*Jadio*	Sangre de grado	*Euphorbiaceae*	Sh	Nu	Y
Juniperus flacida	*Jufla*	Olmo triste	*Cupresaceae*	Sh	Nu	Y
Juniperus deppeana	*Judep*	Táscate	*Cupresaceae*	Tr	Fe-Tu	Y
Juniperus duranguensis	*Judur*	Cedro chino	*Cupresaceae*	Sh	Nu	New
Opuntia leucotricha	*Opuleu*	Nopal duraznillo	*Cactaceae*	Sh	Nu	Y
Opuntia streptacantha	*Opust*	Nopal cardón	*Cactaceae*	Sh	Nu	Y
Prunus serotina	*Pruser*	Cerezo negro	*Rosaceae*	Sh	Ft	Y
Quercus cocolobifolia	*Queco*	Palo manzano	*Fagaceae*	Tr	Fe	Y
Quercus chihuahuensis	*Quechih*	Palo blanco	*Fagaceae*	Tr	Fe	Y

Species	Key	Common name*	Family	Forest classification**	Use¥	Report***
Quercus laeta	*Quela*	Palo blanco	*Fagaceae*	Tr	Fe	Y
Quercus grisea	*Quegri*	Palo chino	*Fagaceae*	Tr	Fe	Y
Quercus potosina	*Quepo*	Palo chaparro	*Fagaceae*	Tr	Fe	Y
Quercus microphylla	*Quemic*	Chaparrito	*Fagaceae*	Sh	Nu	Y
Quercus resinosa	*Queres*	Encino hojudo	*Fagaceae*	Tr	Le	Y
Quercus rugosa	*Querug*	Palo blanco	*Fagaceae*	Tr	To	Y
Quercus sideroxyla	*Quersid*	Palo rojo	*Fagaceae*	Tr	Fe-le	Y
Quercus eduardii	*Queredu*	Palo rojo	*Fagaceae*	Tr	Fe-le	Y
Quercus sp.	*Encino 1*	Encino	*Fagaceae*	Sh	Fe	Y
Quercus obtusata	*Querobt*	Encino	*Fagaceae*	Sh	Fe	New
Yucca filifera	*Yufi*	Palma	*Agavaceae*	Sh	Nu	Y
Odontotrichum amplum	*Adoamp*	Vaquerilla	*Asteraceae*	Sh	Nu	Y
Phytecellobium leptophyllum	*Phylep*	Gatuño de la sierra	*Leguminosae*	Sh	Nu	Y
Eisenhardtia polystachya	*Eipol*	Varaduz	*Fabaceae*	Sh	Nu	Y
Crataegus spp.	*Crasp*	Tejocote	*Rosaceae*	Tr	Nu	New
Quercus sp-2	*Encino 2*	Encino	*Fagaceae*	Sh	Nu	Y
Ipomoea stans	*Ipost*	Galuza	*Convolvulaceae*	Sh	Nu	Y

**The common names were provided by the habitants of "La Congoja" community and do not necessarily correspond to the common name in other localities where these species could be found.*
***Within the forest classification, Tr = Tree and Sh = Shrub.*
****The reports correspond to the flora identified previously, the new reports correspond to the individuals identified in this study.*
¥The use of the forest species recorded, depends on the forest managers experience, in this way, Nu = no use; Fe = firewood extraction; Fr = use as forage plant; Med = Medicinal use; Pt = Timber use; for extraction as fence pole; Ew = extraction as wood stripe from P. teocote; *le = leave extraction for ornamentals; To = tools.*

Table 3.
Forest species identified in an area of the SF-natural protected area, Aguascalientes.

3.1.2 Distribution and abundance of species

The most widely distributed species belong to the genus *Juniperus* (locally known as cedros or táscates), *Quercus* (oaks) and *Arbutus* (locally known as madrone). *J. deppeana* is the most widely distributed species followed by *Quercus potosina* and by *Arctostaphyllos pungens.* The madrones (*Arbutus xalapensis* and *A. glandulosa*) appear in fourth and fifth place, respectively (**Figure** 7).

Out of 50 recorded species, 6 are the ones with the highest abundance indexes. *Q. potosina*, the species best represented in the landscape. This species presents the highest abundance index (ia = 0.1585), followed by *J. deppeana* (ia =0.1102) which also presents the widest distribution. Inside the genus *Pinus*, *P. leiophylla* is the most abundant, even above manzanita (*Arctostaphyllos pungens*) and red oaks (*Q. sideroxyla* y *Q. eduardii*; **Figure 8**).

There are species such as *Pinus chihuahuana*, *Pinus lumholtzii* and *Pinus duranguensis* that present restricted distribution, but are abundant in very specific

Figure 6.
Typical vegetation of the Temperate Mountain Forest. (a) Landscape dominates by conifers in the sierra Fria, in this case, by Pinus leiophylla*; (b) wild ash twig (*Fraxinus uhdei*); (c) Manzanita (*Arctostaphyllos pungens*) specimen in a plateau of the sierra Fria; (d) oak specimen locally known as Palo chino (*Quercus grisea*).*

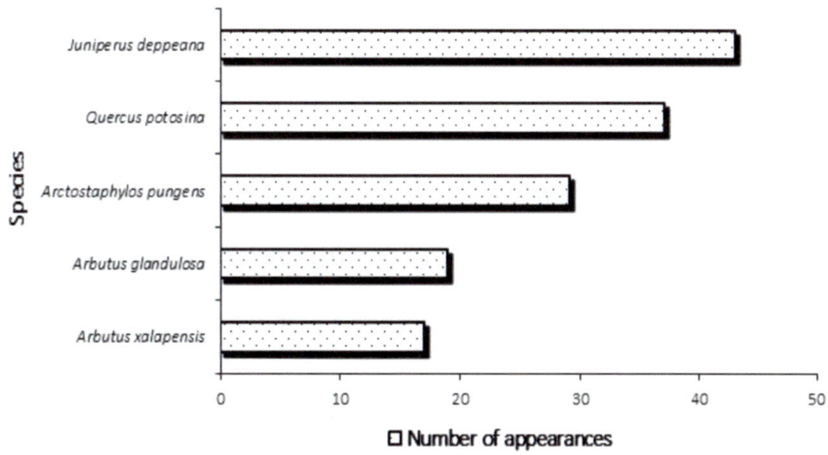

Figure 7.
Forest species with wide distribution in the study area inside the SF-natural protected area.

sites. The distribution analysis based on the altitudinal gradients and geoform suggests that the altitudinal stratum between 2 000 and 2 200 m is the one with the lowest tree and shrub species richness. The best represented species in this range belong to the xeric shrubland being three of them such as *Dodonaea viscosa*, *Phytecellobium leptophyllum*, and *Odontotrichum amplum*, considered as overgrazing indicator species [22]. From the second stratum (2 200 to 2 400 mamsl) *Pinus* and

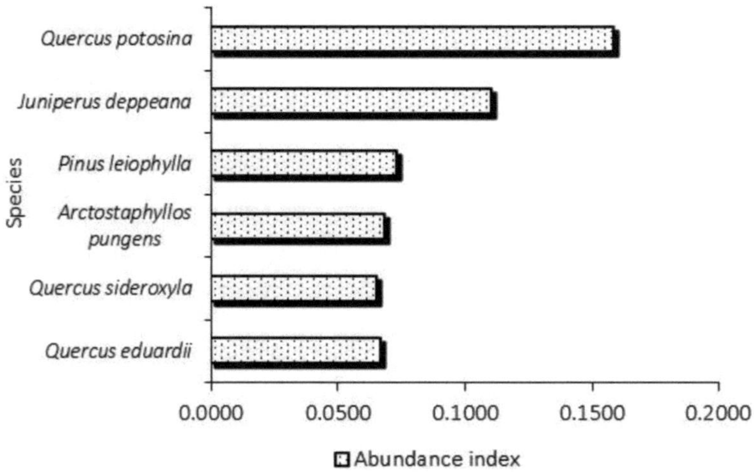

Figure 8.
*Abundancy indexes from the species best represented in the SF-natural protected area, the most representative ecosystem of Temperate Mountain Forest in Ags. The X axis represents the abundancy index which ranges between 0.0670 (*Quercus eduardii*) and 0.1585 (*Q. potosina*). The maximum value of the abundancy index could be 1.*

Quercus species begin to appear, although isolated *Quercus resinosa* individuals can be found at higher altitudes (**Table 4**).

Out of the dominant conifer species at the SF-Natural Protected Area, *Pinus leiophylla* and *P. teocote* are distributed at altitudes ranging from 2 400 to 2 600 masl. Between 2 600 and 2 800 mamsl these two species are more dispersed and located mainly in ravines. *P. leiophylla* is also located on plateaus at 2 700 m (e.g. Mesa del Águila and Mesa del Aserradero). Red oaks (*Q. eduardii* and *Q. sideroxyla*) are distributed at altitudes from 2 400 to 2 600 m, mainly along the ravines (**Table 3**).

In **Figure 9** we shown some species of *Pinus* genera dominants in the intermediate altitudinal strata of SF-Protected Natural Area.

3.2 Dry Tropical Forest

3.2.1 Richness and diversity woody species

We identified 79 species of trees and shrubs, within 45 genera and 14 families (see **Table 5**). The best represented families were Fabaceae (13 genera), Asteraceae (11 genera) and Cactaceae (9 genera). The genera better represented were *Opuntia* (n = 4 spp.), *Acacia* (n = 4 spp.) and *Bursera* (n = 3 spp.). The genero *Salvia* is also important.

The *H′* diversity found in the DTF of the ejido Terrero de la Labor ejido is constant. The highest diversity index found was 3.49 in two of the 26 analyzed sites, which apparently are well conserved sites. On the contrary, three sites had the lowest *H′* diversity index with 2.77 (**Table 6**). Although there are apparently no differences, the highest diversity indexes are located mainly in ravines and north facing exposures, and in locations with difficult access (see **Table** 7).

3.2.2 Distribution and abundance of woody species in the DTF

Of the 79 species identified, eight are distributed in more than 70% of the plots of Terrero de la Labor ejido. The species with the greater distribution are the

	ALTITUDE (MASL)±														
Species	**A1**			**A2**			**A3**			**A4**			**A5**		
	2	**2.1**	**2.19**	**2.2**	**2.3**	**2.39**	**2.4**	**2.5**	**2.59**	**2.6**	**2.7**	**2.79**	**2.8**	**2.9**	**3**
Arctostaphylos pungens															
Dodonaea viscosa															
Juniperus deppeana															
Quercus potosina															
Bursera fagaroides															
Eisenhardtia polystachya															
Juniperus flacida															
Acacia farnesiana															
Prosopis laevigata															
Arbutus glandulosa															
Quercus resinosa															
Yucca filifera															
Phytecellobium leptophyllum															
Asclepias linearis															
Quercus eduardii															
Odontotrichum amplum															
Pinus leiophylla															
Pinus teocote															
Quercus rugosa															
Quercus chihuahuensis															
Quercus sideroxyla															
Arbutus xalapensis															
Pinus lumholtzii															
Juniperus duranguensis															
Quercus cocolobifolia															
Quercus grisea															
Quercus laeta															

**The species distribution in different altitudinal gradients was as a function of the 10 dominant species (obtained from the frequency/site) at each altitudinal stratum.*
±Altitudes (A1-A5) are calculated in m 1000.*
‡The bars with gray shades indicates that this species is abundant at the altitudinal gradient where it was found. In contrast, the black shades indicate that although this species is not abundant, it was found in.

Table 4.
Dominant species distribution by altitudinal strata.

Myrtillocactus geometrizans (garambullo), *Ipomoea murucoides* (palo bobo), *Eysenhardtia polystachya* (varaduz), *Bursera fagaroides* (venadilla), and *Forestiera phillyreoides* (palo blanco) (**Figure 10**), which were located in 96, 92, 90, 88 and 86% of the plots respectively, assuming that the sampling sites are representative of the entire landscape.

On the other extreme, the rarest species were *Plumeria rubra*, *Ficus petiolaris* and *Fraxinus purpurea*. The first species was only located in one site, while the last two

Figure 9.
Populations of Pinus (Pinus *spp) at the SF-natural protected area. The photograph on the left side shows a* Pinus leiophylla *population at the Barranca de Piletas. The pothogragh in the right side shows an image of* Pinus duranguensis. *Photographs as courtesy of Clemente Villalobos llamas and Vicente Díaz Núñez.*

Species	Family	Common name
Acacia berlandieri Benth.	Fabaceae	Carbonera
Acacia farnesiana (L.) Willd.	Fabaceae	Tepame
Acacia pennatula (Schltdl. & Cham.) Benth.	Fabaceae	Huizache o Cascalote
Agave angustifolia Haw.	Asparagaceae	Lechuguilla
Albizia plurijuga (Standl.) Britton & Rose	Fabaceae	Tepeguaje blanco
Alnus acuminata Kunth	Betulaceae	Aile
Amelanchier denticulata (Kunth) K. Koch	Rosaceae	Duraznillo
Amphipterygium molle (Hemsl.) Hemsl. & Rose	Anacardiaceae	Cuachalalate
Asclepias linaria Cav.	Apocynaceae	Algodoncillo
Ayenia mexicana Turcz.	Sterculioideae	
Baccharis heterophylla Kunth	Asteraceae	Escobilla
Bouvardia multiflora (Cav.) Schult. & Schult. f.	Rubiaceae	Clavelito
Brickellia veronicifolia (Kunth) A. Gray	Asteraceae	Orégano de monte
Buddleja cordata Kunth	Buddlejaceae	Tepozán blanco
Buddleja sessiliflora Kunth	Buddlejaceae	Tepozán verde
Bursera bipinnata Donn. Sm.	Burseraceae	Lantrisco
Bursera fagaroides (Kunth) Engl.	Burseraceae	Venadilla
Bursera penicillata (DC.) Engl.	Burseraceae	Arbol de chicle
Calliandra eriophylla Benth.	Fabaceae	Calandria
Castilleja tenuifolia M. Martens & Galeotti	Scrophulariaceae	Hierba del cancer
Cedrela dugesii S. Watson	Meliaceae	Cedro

Species	Family	Common name
Ceiba aesculifolia (Kunth) Britten & Baker f.	Malvaceae	Pochote
Celtis caudata Planch.	Ulmaceae	Capulincillo
Celtis pallida Torr.	Ulmaceae	Vara en cruz
Chusquea sp	Poaceae	Camalote
Colubrina triflora Brongn. Ex G. Don	Rhamnaceae	Algodoncillo
Cordia sonorae Rose	Boraginaceae	Amapa o Vara prieta
Croton ciliatoglandulifer Ortega	Euphorbiaceae	Algodoncillo
Dasylirion acrotrichum (Schiede) Zucc.	Asparagaceae	Sotol
Dodonaea viscosa Jacq.	Sapindaceae	Jarilla
Erythrina flabelliformis Kearney	Fabaceae	Colorín
Eupatorium sp	Asteraceae	Copalillo
Eysenhardtia polystachya (Ortega) Sarg.	Fabaceae	Palo azulo o Varaduz
Eysenhardtia punctata Pennell	Fabaceae	Palo cuate
Ferocactus histrix Lindsay	Cactaceae	Biznaga costillona
Ficus petiolaris Kunth	Moraceae	Ficus silvestre
Forestiera phillyreoides (Benth.) Torr.	Oleaceae	Palo blanco
Fraxinus purpusii Brandegee	Oleaceae	Saucillo
Gymnosperma glutinosum (Spreng.) Less.	Asteraceae	Cola de zorra
Heliocarpus terebinthinaceus (DC.) Hochr.	Meliaceae	Cicuito o Cuero de indio
Ipomoea murucoides Roem. & Schult.	Convolvulaceae	Palo bobo
Iresine sp.	Amaranthaceae	Cola de zorra
Jatropha dioica Sessé	Euphorbiaceae	Sangregrado
Koanophyllon solidaginifolium (A. Gray) R. M. King & H. Rob.	Asteraceae	Caballito
Karwinskia humboldtiana (Schult.) Zucc.	Rhamnaceae	Coyotillo
Leucaena esculenta (Moc. & Sessé ex DC.) Benth.	Fabaceae	Guaje rojo
Lippia inopinata Moldenke	Verbenaceae	Palo oloroso
Lysiloma acapulcense (Kunth) Benth.	Fabaceae	Ébano o Palo fierro Tepeguaje
Lysiloma microphyllum Benth.	Fabaceae	Tepeguaje
Mammillaria bombycina Quehl	Cactaceae	Biznaga de seda
Mammillaria sp.	Cactaceae	Biznaga
Manihot caudata Greenm.	Euphorbiaceae	Pata de gallo
Mimosa monancistra Benth.	Fabaceae	Gatuño o Uña de gato
Mimosa sp.	Fabaceae	Huizache
Myrtillocactus geometrizans (Mart. ex Pfeiff.) Console	Cactaceae	Garambullo
Montanoa leucantha (Lag.) S.F. Blake	Asteraceae	Talacao o Vara blanca
Opuntia leucotricha DC.	Cactaceae	Nopal chaveño o duraznillo
Opuntia robusta J.C. Wendl.	Cactaceae	Tuna tapona
Opuntia sp.	Cactaceae	Nopal
Opuntia streptacantha Lem.	Cactaceae	Nopal cardón

Species	Family	Common name
Perymenium mendezii DC.	Asteraceae	
Pistacia mexicana Kunth	Anacardiaceae	Lantrisco
Pittocaulon filare (McVaugh) H. Rob. & Brettell	Asteraceae	Palo loco
Plumbago pulchella Boiss	Plumbaginaceae	Chilillo medicinal
Plumeria rubra L.	Apocynaceae	Flor de mayo
Prosopis laevigata (Humb. & Bonpl. ex Willd.) M.C. Johnst.	Fabaceae	Mezquite
Ptelea trifoliata L.	Rutaceae	Naranjo agrio o Zorrillo
Quercus laeta Liebm.	Fagaceae	Roble blanco
Salvia mexicana L.	Labiatae	Tlacote
Salvia sp.	Labiatae	Salvias
Stachys coccínea Ortega	Labiatae	Mirto
Stenocereus queretaroensis (F. A. C. Weber) Buxb.	Cactaceae	Pitahaya
Tecoma stans (L.) Juss. ex Kunth	Bignoniaceae	Tronadora
Trixis angustifolia DC.	Asteraceae	Vara verde
Verbesina serrata Cav.	Asteraceae	Vara blanca
Viguiera quinqueradiata (Cav.) A. Gray ex S. Watson	Asteraceae	Vara amarilla
Wimmeria confusa Hemsl.	Celastraceae	Algodoncillo
Yucca filifera Chabaud	Asparagaceae	Palma
Zanthoxylum fagara (L.) Sarg.	Rutaceae	Rabo lagarto

Table 5.
List of species identified in the dry tropical Forest of Terrero de la labor Ejido, Calvillo, Ags.

Altitud level (masl)	Sampled sites	*H´*
1851–1900	7	3.08
101–1950	5	2.57
1951–2000	7	3.60
2001–2100	5	3.14
>2100	2	3.25

Table 6.
Average H' diversity indices associated to different altitudinal ranges in the DTF of Terrero de la labor Ejido.

were only found in two and three sampling sites, respectively. Their low frequency could be associated to their presence in mid statured forests. The most abundant species are those that, even though they are not those with a wide distribution in the landscape, in the places where they are located their frequency is higher than the rest of the identified species. In the DTF of the Terrero de la Labor, the most abundant species belonged to five different genera, of which the most important are *Lysiloma microphylla* (tepeguaje), *Ipomoea murucoides* (palo bobo), and *Bursera fagaroides* (locally known as venadilla) (**Figure** 7). In the case of *Ipomoea murucoides*, it occupies the second place in both distribution and abundance (see **Figure 11**).

Slope range (%)	Sampled sites	*H*′
0–9	6	2.94
9–25	3	2.88
26–37	5	2.92
37–49	4	3.26
49–64	5	3.34
>65	3	3.39

Table 7.
Diversity indexes associated to different slopes of the sites.

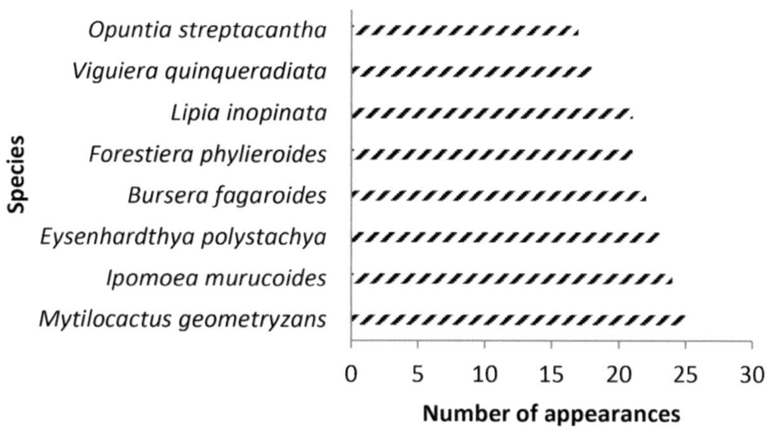

Figure 10.
Species best represented in the DTF Terrero de la labor Ejido.

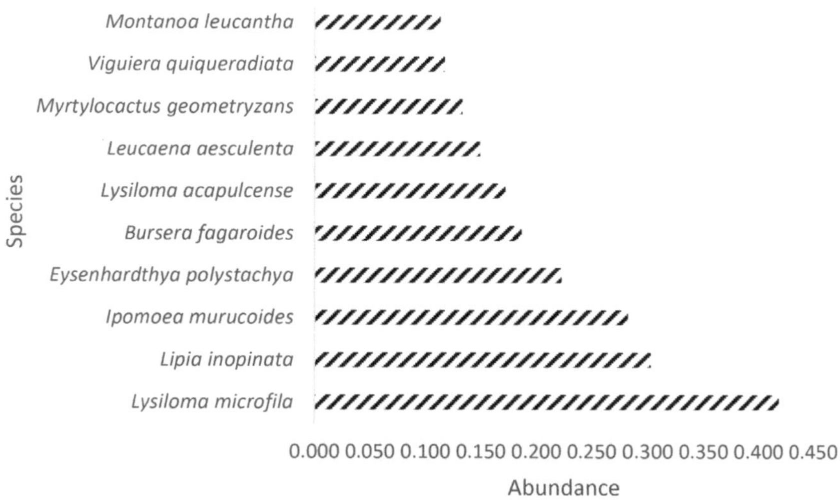

Figure 11.
List of species with the highest abundance in Terrero de la labor Ejido.

The **Figure 12** shown some species of the dominant vegetation in tropical dry forest, in this case, of the Terrero de la labor and las Moras ejidos in the Municipality of Calvillo, Aguascalientes State.

4. Discussion

The loss of biodiversity is one of the environmental problems that has managed to arouse broad global interest in the last two decades [4, 23]. Some of the main

Figure 12.
Diversity of forest species in the tropical dry Forest. (a) Landscape of the tropical dry Forest in the Terrero de la labor and las Moras Ejido; (b) an example of Manihot caudata, *locally known as jaboncillo; (c) specimen of* Bursera fagaroides *(locally known as venadilla or papelillo). Photographs courtesy of Vicente Díaz Núñez, Joaquín Sosa-Ramírez and Jesús Argumedo-Espinoza.*

causes are related to human activities, either directly (overexploitation) or indirectly (habitat alteration), although there is generally an interaction between them. The communication systems have impacted in such a way that both the government and the private sector, as well as society in general, consider a priority to direct greater efforts towards conservation programs. The basis for an objective analysis of biodiversity and its change lies in its correct evaluation and monitoring.

In the Temperate Mountain Forest, the 50 woody species identified show a high species richness in comparison with other mountain regions. The best represented genera correspond to oak trees (*Quercus spp.*) and pines (*Pinus spp.*). The studied area harbors a small portion (6.8%) of the oak species that inhabit Mexico (161 species) [24]; although, this percentage is lower than those reported in areas with a greater territorial surface and higher rainfall, such as the case of San Luis Potosi and Jalisco States, which has identified 45 and 51 oak species respectively, that represent 27.95 and 36.9% of the total oak species registered in Mexico [25], the SF-NPA represent less 5% of the territorial surface in the mentioned states. In relation to pines, the studied area has about 17% of the species identified in Mexico [26]. This proportion is similar to that reported by Márquez-Linares et al., [27] in an area of pine-oak forest, in Durango, Mexico, where they recorded 8 pine species. In relation to "Las Joyas" scientific station, in the Sierra de Manantlán Biosphere Reserve, the *Quercus* diversity (16 spp.) is similar to the one in the Sierra Fria, although the area of las Joyas is smaller (Ca. 3600 ha.).

In the Sierra Fría, the most widely distributed and abundant species are the potosine oak (*Q. potosina*) and alligator juniper (*J. deppeana*). In the case of *Q. potosina*, its distribution and abundance may be related to the dominant physiography in this area, as well as to the mean annual precipitation (650 mm). The appearance of *J. deppeana* is possibly related to the disturbances that occurred in the Sierra Fria during the period between 1920 and 1940 [28]. This species has probably been a pioneer in the recovery of the vegetation cover, although the presence of manzanita (*A. pungens*) has also been documented colonizing sites where disturbances occurred, either natural, as in the case of fires or, anthropogenic, such as forest clearance and harvesting. Pines population is restricted to the Sierra Fría and the Sierra del Laurel. In the Sierra Fria, *Pinus teocote* (locally known as pino ocote) and *Pinus leiophylla* (locally known as pino prieto) are the two most abundant pine species. Its population is abundant in humid places and altitudes higher than 2,500 masl. *P. leiophylla* isolated specimens have been found on flat lands, which suggests that in the past this species had a greater distribution. In the Sierra del Laurel only two pine species have been identified, the pino triste (*Pinus lumholtzii*) and the pino piñonero (*Pinus cembroides var. cembroides*) in isolated populations, which suggests that in the past they were more abundant; however, the existing information is incipient.

The *H'*diversity indexes for each altitudinal stratum suggest that, between 2,400 and 2,600 mamsl, the plant richness of the SF-Natural Protected Area is similar to temperate forests, similar to what the β Whitakker index showed.

The distribution of species such as *J. deppeana* and *Q. potosina*, the most abundant and widely distributed, are influenced by flat sites and canopy covers that vary between 30 and 50%. One explanation is that *Q. potosina* tolerates high drought rates and *J. deppeana* is a pioneer species in disturbed sites, as suggested by Minnich et al. (1994) [28]. On the other hand, the presence or absence of the species may also be dictated due to their dispersal capacity or to the presence or absence of dispersers [19]. The results obtained contribute to describe the habitat of the species, which is an essential factor in programs aiming the restoration and management of temperate climate forests [8, 29], actions that, at least in the case of Mexico, have shown few results.

The species richness in BTS is generally lower than in humid tropical forests [30], although higher than in Temperate Mountain Forests [25]. The BTS is dominated by relatively short trees, most of which lose all their foliage during the dry season. In this community, herbaceous life form, thin woody species, and vines are common, but epiphytes and thick lianas are less abundant and diverse than in humid forests [31]. Diversity is generally higher without a clear dominance of any species, to the point that many of them are rare [32]. In this type of ecosystem, it is common to identify some genera such as *Bursera, Lonchocarpus, Lysiloma* and *Jatropha*, as well as emerging columnar cacti [33].

The species richness found at the Terrero de la labor Ejido BTS (N = 79) is similar to that reported by Trejo (2005) [33], where he points out that on average the tropical dry forest in Mexico harbors around 74 species with a DBH ≥ 1 cm in 0.1 ha. However, in the study site, some species considered "rare" which are indicators of medium forest (e.g *Amphipterygium molle*) were found in ravines and better preserved sites, suggesting that at some point this ecosystem had a greater presence in the landscape.

The analysis of the diversity, distribution and abundance associated with the Tropical Dry Forest in Aguascalientes has been little addressed, so the study conducted in the BTS of the Municipality of Calvillo represents one of the first efforts to understand this ecosystem natural heritage [14]. Previously, partial floristic studies had been carried out, studies which mainly referred to the dominant vegetation types and some important species, however on these studies there were gaps in relation to the ecology of the plant communities [20]. On the other hand, other studies mention some factors related to the mortality of these natural communities [15], but there is no information on vegetation diversity which reflects the real tropical dry forest importance.

This work contributes directly to the management of the ecosystems analyzed. Knowledge about species richness and their distribution provides an overview of the territory's conservation state, considering that both the Temperate Mountain Forest and the Tropical Dry Forest studied are part of the Sierra Fria Protected Natural Area, which is the protected area with the biggest extension in the State. On the other hand, the bases are established for the restoration of degraded ecosystems, either through active restoration or through mechanisms of ecological succession (passive restoration) [29].

Acknowledgements

The authors acknowledge the participation of Jesus Argumedo-Espinoza for his cartographic support. Likewise, we thank the facilities provided of the owners of the Sierra Fria, as well as Jesus Velasco Serna of the Terrero de la Labor ejido for in the gathering of field information.

Author details

Joaquín Sosa-Ramírez[1]*, Vicente Díaz-Núñez[2] and Diego R. Pérez-Salicrup[3]

1 Agricultural Sciences Center, Autonomous University of Aguascalientes, Av. Universidad, Ciudad Universitaria, Aguascalientes, Ags., México

2 Scientific Research Invited by the Agricultural Sciences Center, Autonomous University of Aguascalientes, Avenida Universidad, Ciudad Universitaria, Aguascalientes, Ags., México

3 Ecosystem Research Institute, National Autonomous University of Mexico, Morelia, Mich, Mexico

*Address all correspondence to: jsosar@correo.uaa.mx

References

[1] Sarukhán, J., P. Koleff, J. Carabias, J. Soberón, R. Dirzo, J. Llorrente-Busquets, G. Halfter, R. González, I. March, A. Mohar, S. Anta, J. de la Maza. 2009. Natural Capital of Mexico: current knowledge, evaluation and perspectives of sustainability. Synthesis/Capital Natural de México: conocimiento actual, evaluación y perspectivas de sustentabilidad. Síntesis. México: CONABIO. 100 p.

[2] Llorente-Bousquets, J., y S. Ocegueda. 2008. State of knowledge of biota. In: CONABIO (2008). Natural Capital of Mexico, vol. I: Current knowledge of biodiversity/Estado del conocimiento de la biota. En: CONABIO (2008). Capital natural de México, vol. I: Conocimiento actual de la biodiversidad. CONABIO, México, 2008.

[3] Mora, F. 2019. The use of ecological integrity indicators within the natural capital index framework: The ecological and economic value of the remnant natural capital of Mexico. Journal for Nature Conservation 47:77–92

[4] Organización de las Naciones Unidas para la Agricultura y la Alimentación (FAO). 2016. Global Forest Resources Assessment 2015: How are the world's changing? Second edition. FAO. 54 p. Available in: http://www.fao.org/3/a-i4793e.pdf

[5] Organizaciòn de las Naciones Unidas para la Agricultura y la Alimentación (FAO). 2010. Evaluación de los Recursos Forestales Mundiales. Informe Principal/Global Forest Resources Assessment. Main Report. FAO. Roma. 381 p.

[6] Challenger, A. y J. Soberón. 2008. Los ecosistemas terrestres de México. En: Capital natural de México, Vol. I: Conocimiento actual de la biodiversidad, J. Soberón, G. Halfter y J. Llorente (eds.)/Terrestrial ecosystems of Mexico. In: Natural Capital of Mexico, Vol. I: Current knowledge of biodiversity, J. Soberón, G. Halfter and J. Llorente (eds.). Comisión Nacional para el Conocimiento de la Biodiversidad –CONABIO-, México. ISBN 978–607-7607-03-8. p. 87–108.

[7] Balvanera, P. 2012. Los servicios ecosistèmicos que ofrecen los bosques tropicales/ The ecosystem services offered by tropical forests. Ecosistemas 21: (1). http://www.revistaecosistemas.net/articulo.asp?Id=709

[8] Baskent, E.Z. 2020. A framework for characterizing and regulating ecosystem services in a management planning context. Forests 11: 102; doi:10.3390/f11010102

[9] Comisión Nacional Forestal (CONAFOR). 2014. National Forest and Soil Inventory 2014–2018. Main report/ Inventario Nacional Forestal y de Suelos 2014–2018. Informe de resultados. SEMARNAT-CONAFOR. 200 p.

[10] Díaz, V., J. Sosa-Ramírez y D.R. Pérez-Salicrup. 2012. Trees and shrubs Distribution and abundance in Sierra Fria, Aguascalientes, Mexico/ Distribución y abundancia de las especies arbóreas y arbustivas en la Sierra Fría, Aguascalientes, México/. Polibotánica 34: 99–126.

[11] Inventario Estatal Forestal y de Suelos (IEFyS). 2012. State Forest and Soil Inventory of Aguascalientes/ Inventario Estatal Forestal y de Suelos de Aguascalientes. Gobierno del Estado de Aguascalientes-SEMARNAT-CONAFOR. 122 P.

[12] Gobierno del Estado de Aguascalientes. 2015. Decree of the Protected Natural Area "Área Silvestre Estatal Sierra Frìa"/Decreto del Área Natural Protegida "Área Silvestre Estatal

Sierra Fría". Diario Oficial del Estado de Aguascalientes. Disponible en: http://eservicios2.aguascalientes.gob.mx/NormatecaAdministrador/archivos/EDO-12-47.pdf

[13] Díaz-Núñez, V., J. Sosa-Ramírez, and D.R. Pérez-Salicrup. 2016. Vegetation patch dynamics and tree diversity in a conifer and oak forests in Central Mexico. Botanical Sciences 94: 229–240.

[14] Argumedo-Espinoza, J., J. Sosa-Ramírez, V. Díaz-Núñez, D.R. Pérez-Salicrup, and M.E. Siqueiros-Delgado. 2017. Diversity, distribution and abundance of woody plants in a dry tropical forest: recomendation for its management. In: Ortega-Rubio (Ed). 2017. Mexican Natural Resources Management and Biodiversity Conservation. Springer. Pp. 479–500

[15] Díaz-Núñez, V., J. Sosa-Ramírez e I. P. Macías-Medina. 2014. Phytosanitary diagnosis of vegetation in priority ecosystems of Aguascalientes, Mexico/ Diagnóstico fitosanitario de la vegetación en ecosistemas prioritarios de Aguascalientes, México. Comisión Nacional Forestal-Secretaría de Medio Ambiente del Estado de Aguascalientes. 84 p.

[16] Daget Ph. y M. Godron. 1982. Analysis of the ecology of species in communities/ Analyse de l'ecologie des espéces dans les communautés. Masson, Paris. 163p.

[17] De la Cerda, M. E. 1999. Oaks of Aguascalientes/Encinos de Aguascalientes. Universidad Autónoma de Aguascalientes. Segunda Edición. 77 p.

[18] Siquéiros, D.M.E. 1989. Conifers of Aguascalientes/Coníferas de Aguascalientes. Universidad Autónoma de Aguascalientes, 68 p.

[19] Krebs, C. K. 1993. Factors that limit distributions: Dispersal. In Ecology, C. K. Krebs. John Whiley and Sons. p. 41–56.

[20] Siqueiros-Delgado, M.E., J.A. Rodríguez-Ávalos, J. Martínez-Ramírez y J.C. Sierra-Muñoz. 2016. Current status of the vegetation of Aguascalientes, Mexico/ Situación actual de la vegetación del Estado de Aguascalientes/. Botanical Scienses 94: 455–470

[21] INEGI. 2007. Agrarian nuclei. Municipality basic tables. Certification Program for Ejido Rights and Land Titling/Núcleos agrarios. Tabulados básicos por municipio. Programa de Certificación de Derechos Ejidales y Titulación de Solares. PROCEDE. México.

[22] Begon, M., C. T. Towsend y J. L. Harper. 2006. Ecology: from individuals to ecosystems. Blackwell Publishing, Oxford. 738 p.

[23] Brawman, K.A., and G.C. Daily. 2008. Ecosystem services. Human ecology. Elsevier. Pp. 1148–1154

[24] Valencia, A. S. 2004. Diversity of *Quercus* genera (Fagaceae) in Mexico/ Diversidad del género *Quercus* (Fagaceae) en México. Boletín de la Sociedad Botánica de México 75: 33–53.

[25] Sabás-Rosales, J.L., J. Sosa-Ramírez, and J.J. Luna-Ruiz. 2015. Diversity, distribution and basic habitat characterization of the Oaks (*Quercus*: Fagaceae) of San Luis Potosí, México/ Diversidad, distribución y caracterización básica del hábitat de los encinos (*Quercus*: Fagaceae) del Estado de San Luis Potosí, México. Botanical Sciences 93: 881–897. DOI 10.17129/botsci.205

[26] Saénz-Romero, C., A. E. Snively y R. Lindig-Cisneros 2003. Conservation and restoration of pine forest genetic resources in Mexico. Silvae Genetica 52: 233–237.

[27] Márquez-Linares, M. A., S. González-Elizondo y R. Álvarez-Zagoya. 1999. Components of tree diversity in pine-oak forests of Durango, Mex./ Componentes de la diversidad arbórea en bosques de pino encino de Durango, Mex. Madera y Bosques 5: 67–78.

[28] Minnich, R. A., J. Sosa-Ramírez, V. E. Franco, W. J. Barry y M. E. Siqueiros. 1994. Preliminar recognizing of vegetation and human impacts in Sierra Fria, Aguascalientes/Reconocimiento preliminar de la vegetación y de los impactos de las actividades humanas en la Sierra Fría, Aguascalientes. Investigación y Ciencia 12: 23–29.

[29] Rey-Benayas, J. M., J. M. Bullock y A. C. Newton. 2008. Creating woodland islets to reconcile ecological restoration, conservation, and agricultural land use. Frontiers in Ecology and the Environment 6: 329–336.

[30] Gentry, A. H. 1995. Diversity and floristic composition of neotropical dry forests. In Seasonally dry tropical forests, S. H. Bullock, H. A. Mooney y E. Medina (eds.) Cambridge University Press, Cambridge. p. 146–190.

[31] Pineda-García, F., L. Arredondo-Amezcua y G. Ibarra-Manríquez. 2007. Richness and diversity of wood species in deciduous tropical forest of El Tarimo, basin of del Balsas, Guerrero/ Riqueza y diversidad de especies leñosas del bosque tropical caducifolio El Tarimo, Cuenca del Balsas, Guerrero/ Richness and diversity of Woody species in the tropical dry forests of El Tarimo, Cuenca del Balsas, Guerrero. Revista Mexicana de Biodiversidad 78: 129–139.

[32] Durán, Z.V.H., M.J.R. Francia, P.C. R. Rodríguez, R.A. Martínez y R.B. Cárceles. 2006. Soil-erosion and runoff prevention by plant covers in a mountainous area (SE Spain): Implications for sustainable agriculture. The Environmentalist 26:309–31.

[33] Trejo, I. 2005. Analysis of the diversity of the deciduous tropical forest in Mexico. In: Halfter et al., (2005). On biological diversity: the meaning of alpha, beta and gamma diversities/ Análisis de la diversidad de la selva baja caducifolia en México. En: Halfter et al., (2005). Sobre diversidad biológica: el significado de las diversidades alfa, beta y gama. CONABIO, SEA, CONACyT. ISBN 84–932807–7-1. Pp. 111–122.

Chapter 3

Tropical Subterranean Ecosystems in Mexico, Guatemala and Belize: A Review of Aquatic Biodiversity and Their Ecological Aspects

Luis M. Mejía-Ortíz, Peter Sprouse, Juan C. Tejeda-Mazariegos, Jair Valladarez, Oscar Frausto-Martínez, Alejandro L. Collantes-Chavez-Costa, Guillermo Ruíz-Cancino and German Yáñez

Abstract

The subterranean ecosystems in tropical areas of Mexico, North of Guatemala & Belize are very abundant because the karstic soil that allow these formations are the main composition in the Yucatán Peninsula and several mountains systems in these countries; also, they have a strong relationship with tropical forest adjacent where the main energy into the caves have an alloctonous origin. In these three countries there are three different cave conditions: a) freshwater semi-dry caves, b) flooded freshwater systems and c) anchialine systems. Mainly crustaceans and freshwater fishes are the major representative group in the aquatic diversity in these systems because the anchialine members are restricted to Yucatán Peninsula and Islands adjacent. Around 5000 entries to subterranean world there are among these countries, where the Yucatan Peninsula is the area with major caves or cenotes in comparison with southern of Mexico, North of Guatemala and Belize. Into these systems are possible found crustaceans and fishes from different families. The objective of this paper is present a review of these systems according with each karstic areas and show the current map including the location of each systems; as well their subterranean aquatic biodiversity and, finally discuss the relationships among these different areas using their biological aquatic richness in consideration with ecological subterranean conditions.

Keywords: Mesoamerica, Subterranean Biodiversity, Cave environments

1. Introduction

The biodiversity in the tropical area among Mexico, Guatemala and Belize are good represented in several taxa since the tropical forest are the most representative biomass in this area, where there are a high species richness. The climatic conditions produce a great opportunity to maintain this diversity [1].

The geological history of Central America shows that this area is recently in comparison with the Mexican North portion, however the Peninsula Block has moved into the sea 165 Ma during the Jurassic Period, and emerge on Pleistocene; during these last period the opportunity of the migration species to colonised the mountain chains in Guatemala y Belize had origin from north to south [2].

This geological history include the karstic regions in the three countries, but the Yucatan Peninsula is the most recently portion in emerge from the sea in México [3], but the volcanic activities and emerged areas in the north of Guatemala and even Belize central portion involved the mountain formation and limestone soil, and of course the colonised caves and grootes from terrestrial and aquatic animals from surface involved species that in another times previously had occupied these epigeal environments [4, 5]. The geological conditions produce a different opportunities to be occupied for these animals mainly crustaceans and fishes [6, 7].

Has been reported the existence of different aquatic habitats in the cave environments: a) freshwater semi-dry caves, b) flooded freshwater systems and c) anchialine systems, only Guatemala have not the anchialine systems reported, but the three conditions are present in Mexico & Belize [8–16], to date is possible identify four main karstic areas: Chiapas Mountains, Yucatan Peninsula (Mexico), Alta VeraPaz karstic area, Peten Area (Guatemala), and Chiquibul area (Belize). Where fishes and crustaceans has been reported with several species. The aim of present chapter is show a review of the aquatic subterranean biodiversity in these two major taxa and their relationship with their ecological conditions that there are in each habitat type, to discuss the interrelationships that these five karstic areas.

2. Study area

The study area includes the karstic regions from Tehuantepec Isthmus, Yucatán Peninsula and Islands adjacent and Guatemala & Belize. In this area exist five karstic regions 1) Chiapas karstic Mountains, 2) Yucatán Peninsula, 3) Alta Verapaz, 4) Petén Area, and 5) Chiquibul area as is possible see in the **Figure 1**.

In each area has been recorded several entrance to subterranean environments in the all them the dissolution caves are the principal formation due the limestone soil. The Chiapas karstic mountains had diverse caves record mainly dry and semi-dry conditions but in there are the most origins of superficial springs that flow to Gulf of Mexico basin [8, 9, 17]. In the Yucatán Peninsula is where there are more entrance recorded due the special efforts that involve the anchihaline systems but also there are an important number of dry cave or semi-dry caves in there in special on the Riviera Maya Coast. Also show a sub-region called Cenotes Ring where the freshwater flooded caves is very well represented [18]. Also in the Island and coast as Cozumel Island there are anchialine caves with species fauna [12–14]. Alta Verapaz region had an important Mountain Chain that produce conditions to dissolution of limestone that is the main characteristic of soil to produce different cave types from Springs as Hunalye and semi-dry caves Lanquin area where some freshwater prawns, crabs, and fishes living [19]. The highlands have an estimation of 150 Ma of age. The Petén region is the Peninsula base where the limestone starts to be the main soil type and the cave formations are frequently from Quaternary times with marine sediments [5]. Finally the Chiquibul area in the Mountain Central part of Belize but as a continuum of these soil, show a unexplored area even to discover, however in the area at less there are some caves records with important fauna [11].

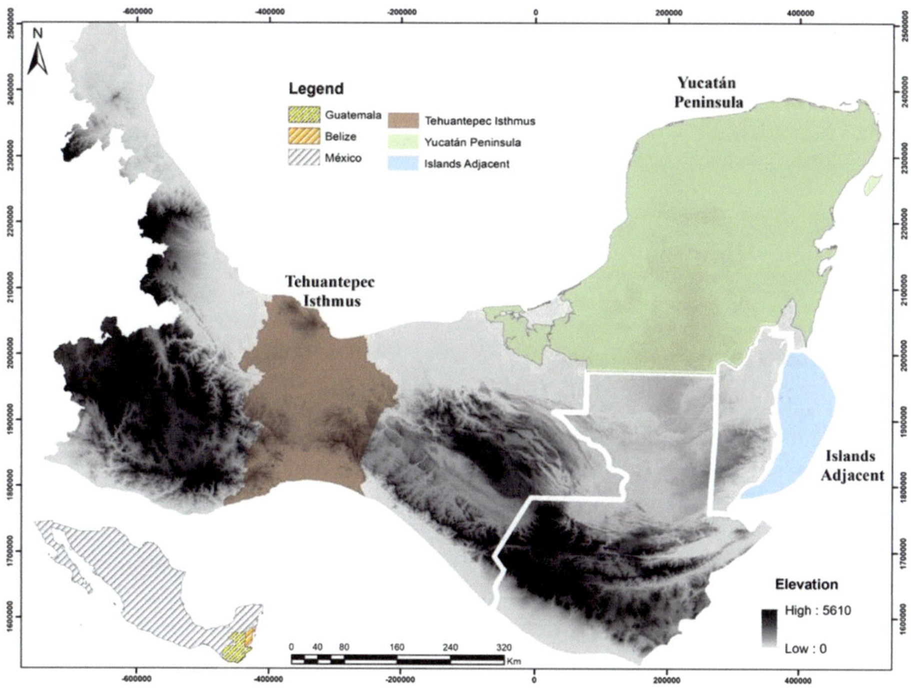

Figure 1.
Mesoamerican karstic region. Include the five study areas: a) isthmus and Chiapas Mountain systems, b) Yucatan peninsula, c) Alta Verapaz, d) Peten & e) Chiquibul and karst Main Mountain.

3. Material & methods

Several explored trips where conducted on the last fifteen years around the study area, exploring caves and cenotes in each karstic region; according with the site was measured the abiotic data from the water (temperature, conductivity, salinity, pH, depth, dissolved oxygen and light) using the Hydrolab Data Sonde 5, applying SCUBA techniques; or the oximeter Oakton: dissolved oxygen ($\pm$0.01 mg/l), pH ($\pm$001 pH), salinity ($\pm$0.01%), and temperature of the water ($\pm$0.01°C). The altitude and the GPS values were recorded with a Garmin GPS [20–24]. With help to several speleological groups were record the GPS data from each entrance, and the photographic record of fauna has been registered. More than 100 caves were visited as representative from all regions were recorded fishes and crustaceans and preserved in alcohol to taxonomic identification, and compared with previous reports. The maps of entrance distribution to each region were done and the fauna richness relationship analysis was made according with the procedure reported for some subterranean systems [25], that "due the possibility to have a samples incomplete, some estimators have been derived to predict the true number of species based on rare species in a sample. This was calculate according with Chao [26],

$$S_1 = S_{obs} + \left(L^2/2M\right) \tag{1}$$

where S_{obs} is equal to the number of species observed in a sample, L is the number of observed species represented by a single individual (i.e., singletons), and M is the number of observed species represented by two individuals in the sample (i.e., doubletons).

Also the same authors recommend the application of Burnham and Overton's [27] jackknife estimators in order to reduce estimation bias in estimating species richness.

$$S_2 = S_{obs} + \left[(L(2n-3)/n) - \left(M(n-2)^2/n(n-1)\right)\right] \quad (2)$$

where n is the number of samples. No direct formula for the calculation of the variance is available".

4. Results

4.1 Karstic areas and their aquatic diversity

a. Chiapas Mountains (**Figure 2**)

The Chiapas Karstic area involve the Mountain Chain from Tehuantepec Isthmus to Guatemala border, in there these mountains have a karstic soil and the dissolution of limestone produce several caves. The altitude range is from 1700 to 100 meters above sea level (masl) and in these regions there are almost 150 caves reported. The drainage from aquifer that flow to Gulf of Mexico produce one connection with surface and many cave crustaceans had their origins from epigeal populations. The main fishes in this region are from *Rhamdia* genus with or without adaptation to cave life. In the crustaceans decapods species there are freshwater prawns *Macrobrachium sbordonii*, *Macrobrachium acherontium*, *Cryphiops sbordonii* and *Cryphiops luscus*; crayfishes *Procambarus mirandai*, *Procambarus* sp. (From La Lucha system)

Figure 2.
Caves location from isthmus and Chiapas region. The red points represent each entrance to subterranean systems.

crabs *Rodriguezia adani*, *Avotrichodactylus bidens*, and *Rodriguezia* spp. (From the Ocosingo Area) (**Table 1**). In this region only the freshwater crabs was recorder from literature the remain was confirmed with our fieldtrip work.

Crustaceans	Fishes
Procambarus mirandai	*Rhamdia guatemalensis* -Cosmopolitan species
Procambarus sp. (Itshmus)	*Rhamdia sbordonii*
Macrobrachium acherontium	
Macrobrachium sbordonii	
Macrobrachium sp. (Isthmus)	
Cryphiops sbordonii	
Cryphiops luscus	
Rodriguezia adani	
Avotrichodactylus bidens	
Typhlopseudothelphusa mocinoi	
Typhlopseudothelphusa hyba	
Rodriguezia villalobosi	
Rodriguezia mensabak	
Odontothelphusa monodontis	

Table 1.
Checklist of subterranean fauna taxa of the Tehuantepec isthmus and Chiapas region all from freshwater semi-dry caves.

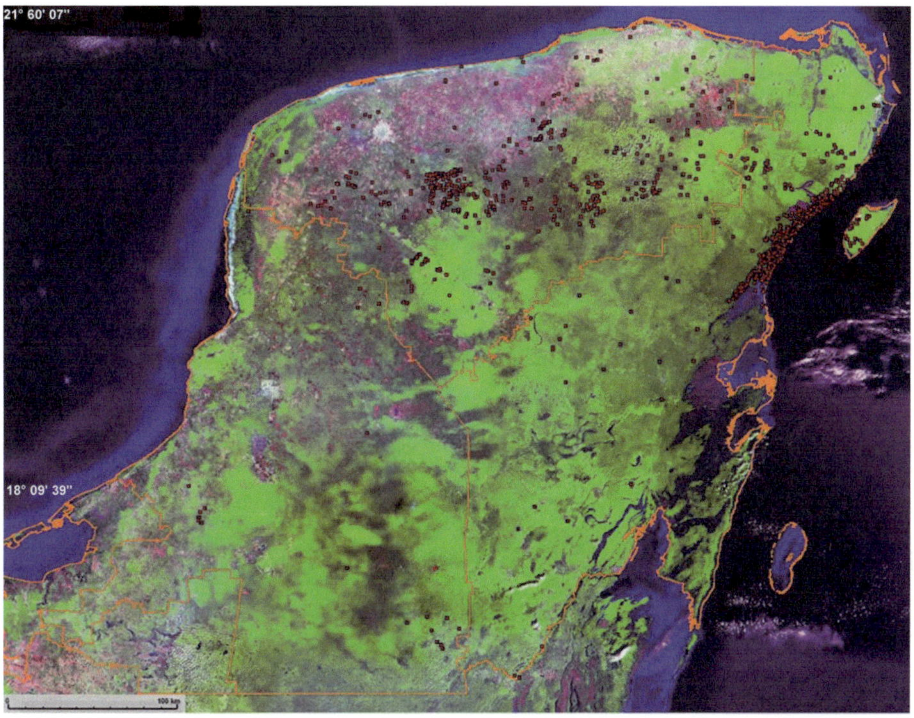

Figure 3.
Caves and cenotes from Yucatan peninsula. The red points represent each entrance to subterranean systems.

b. Yucatan Peninsula (**Figure 3**)

This area involve three Mexican states, is the longest area in the Southern of Mexico with a major caves formations involve the semi-dry, dry and flooded caves, in those closed to coast with sea water and freshwater subterranean interactions are called anchihaline systems, and their fauna is complete different to those with all freshwater. In the Yucatan Peninsula the maximum elevation is in Ticul Mountains, with 300 masl, However according with the different geological times that this Peninsula emerge there are at less five sub-regional areas, where the tropical forest are the most source of energy to maintenance the live in underground. The fishes more representative in the region are catfishes, *Rhamdia guatemalensis*, *Opisternon infernale* and *Ogilbia pearsei*, all them in freshwater, in the same conditions the crustaceans more abundant are *Creaseriella anops* (isopod), *Creaseria morleyi* and *Typhlatya mitchelli* and *Typhlatya pearsei*. Whilst, in anchialine systems there are *Barbouria cubensis, Agostocaris bozanici, Agostocaris zabaletai, Anchialocaris paulini, Procaris mexicana, Parhippolyte stereri, Yagerocaris cozumel, Xibalbanus tulumensis, Xibalbanus cozumelensis* and *Xibalbanus fuchscockborni,*

Freshwater habitats	
Crustaceans	**Fishes**
Creaseriella anops (Isopod) Cosmopolitan species	*Rhamdia guatemalensis* –Cosmopolitan species
Creaseria morleyi –Cosmopolitan species	*Opisternon infernale*
Typhlatya mitchelli –Cosmopolitan species	*Ogilbia pearsei*
Typhlatya campechae	
Procambarus sp.	
Anchialine habitats	
Barbouria cubensis	
Anchialocaris paulini	
Agostocaris bozanici	
Agostocaris zabaletai	
Yagerocaris cozumel	
Triacanthoneus akumalensis	
Parhippolyte sterreri	
Janicea antiguensis	
Calliasmata nohochi	
Procaris mexicana	
Typhlatya dzilamensis	
Typhlatya pearsei	
Xibalbanus cozumelensis	
Xibalbanus tulumensis	
Xibalbanus fuchscockborni	
Metacirolana mayana	

Table 2.
Checklist of subterranean fauna taxa of the Yucatan peninsula region.

Metacirolana mayana, Mayaweckelia cernua (**Table 2**). In this case all crustaceans was confirmed with our fieldtrip work.

c. Alta VeraPaz Region (**Figure 4**)

This area comprised the Mountains Chains that slope drainage to Gulf of Mexico, there are a continuum of Mountains from Chiapas, and they are formed mainly by karstic soil, and the elevations go to 2000 m from Río Salinas and Río Xcán, that drainage to Usumacinta river in Mexico and to Cahabon river that go to Izabal Lake. In they are the main cave formations are in the Lanquin Area, and the springs from different rivers such the Hunalye, Cahabon, Xcán and others. The species reported in there are: freshwater prawns *Macrobrachium vicconi* in the entrance of spring of Hunalye without cave adaptations, *Macrobrachium* spp. (in description process), blind crabs, from Pseudothelphusidae family, and catfishes from *Rhamdia* genus. The tropical forest is the most common adjacent ecosystems and the bat activity to carried energy inside the caves is the principal source to maintenance the ecological function from these underground ecosystems. In this case all species was collected by authors.

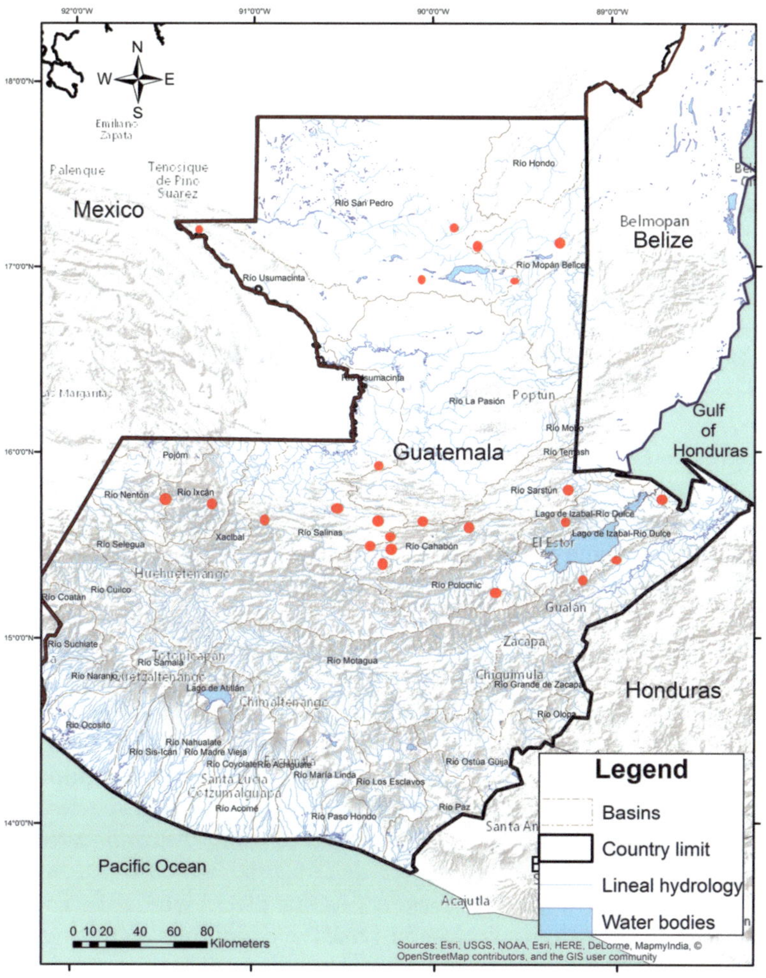

Figure 4.
Caves from Alta Verapaz and Peten regions in Guatemala.- the red points represent each entrance to subterranean systems.

d. Petén Region (**Figure 4**)

This area is part of the Yucatan Peninsula base, where the elevations are less evident, from 350 to 50 masl, and the drainage to underground by karstic soil of water is more representative, here the caves are on floor level, and local people as water source use them. In this area the principal species are freshwater prawns *Macrobrachium* and crabs from Pseudothelphusidae family. They are not showed cave life adaptations. In this region all animals was confirmed by fieldwork from authors.

e. Chiquibul Region (**Figure 5**)

In Belize, there are several areas with karstic composition in the soil but only around of Mountains Systems has been recorded caves with long formations. Is important mentioned that close to Mexican border the soils is too similar to rest of Yucatan Peninsula and the potential to found caves or even cenotes is high. However, the caves recorded in Belize are mainly in Chiquibul region among 150 to 800 msal and in Islands and Cays where has been recorded anchialine systems. In there some crustaceans could be found as *Macrobrachium catonium*, *Typhopseudothelphusa acanthochela* and fishes from *Rhamdia laticauda*, however there are in description two new species of *Macrobrachium* species and one crab from the same family. In the anchialine systems has been reported *Xibalbanus cockei.* Jill Yager author that described it confirmed only this last species.

4.2 Ecological conditions to freshwater and anchialine habitats

The geological history of this region has two main sections in first instance the Mountain Systems Development in the different geological times was producing a new subterranean habitats to some freshwater groups invaded these sites with a consequently a new opportunity to speciation but they could be seen as a biological subterranean corridor, because these species have the same selection pressures and the changes among them are so closed. However, as has been reported each cave is a new chance to produce some changes in the adaptation as outcome of isolation procedure that considering this the different species of freshwater prawns, crayfish, or crabs, in these places in average the oxygen are lower between 2 to 3 mg/l with a saturation of 60%; at same time the pH is around the neutral values with some peaks to alkalinity, normally all with freshwater conditions and values of temperature around the 18 to 22°C, in all these places the measures were taken with low depth.

However, the Yucatan Peninsula have a different origin and the different ways to colonised this subterranean habitats, our results show that in the enormous plate the species are cosmopolitan but exclusively in they are as *Creaseria morleyi, Typhlatya mitchelli*. Another big faunistic group is from anchialine group that their marine habitats conditions there are species so very primitive as Remipedia with at less three species around the coastal caves, or different decapod species that has been reported with different origins, as *Procaris mexicana, Anchialocaris paulini, Agostocaris bozanici, A. zabaletai, Typhlatya dzilamensis, Barbouria cubensis*, or even *Calliasmata nohochi* and *Yagerocaris cozumel*. In this places where the anchialine habitats are present the salinity is closed to marine conditions 36 ups, with pH values clearly to alkalinity between 8 and 9, the oxygen dissolved are close to hypoxic conditions 0.15 to 0.3 mg/l, with 5 or 10% of saturation. The temperature is around the 24°C [28] an example of these behaviour is showed in the **Figure 6** to Cenote Chempita.

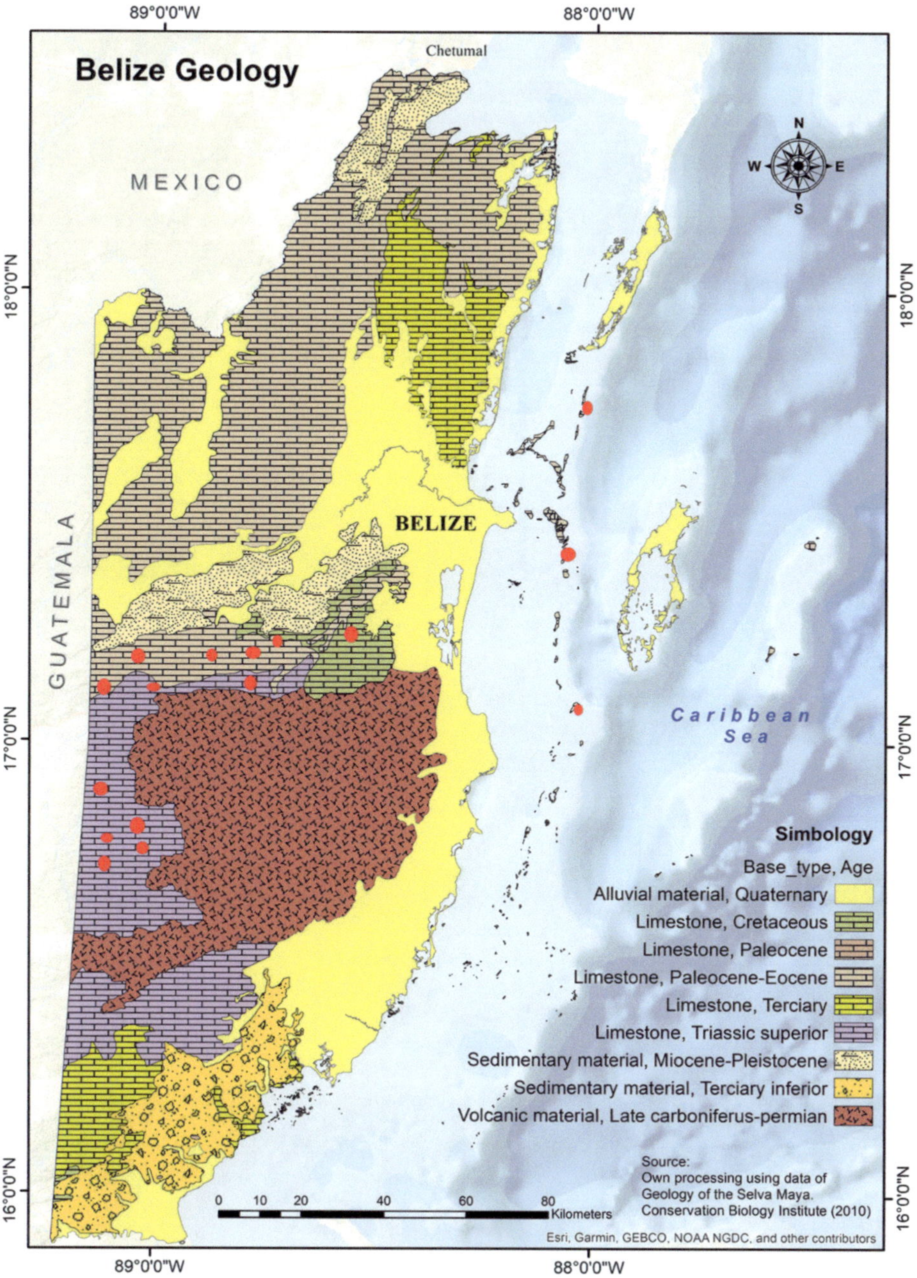

Figure 5.
Caves from Belize karst regions. The red points represent each entrance to subterranean systems.

4.3 Interrelationships among the karstic areas (richness analysis and biological subterranean corridor)

The five karstic areas involved, show a differences in the species composition and their numbers, in some places only one specimens are located, whilst another there are hundred or even miles (**Table 3**). However, in the first view or richness the Yucatan Peninsula is more diversity in the crustaceans but have two different habitats and in the freshwater the *Creaseria morleyi* and *Creaseriella anops* are too cosmopolitan; whilst in the coastal areas the anchialine habitats allow that species

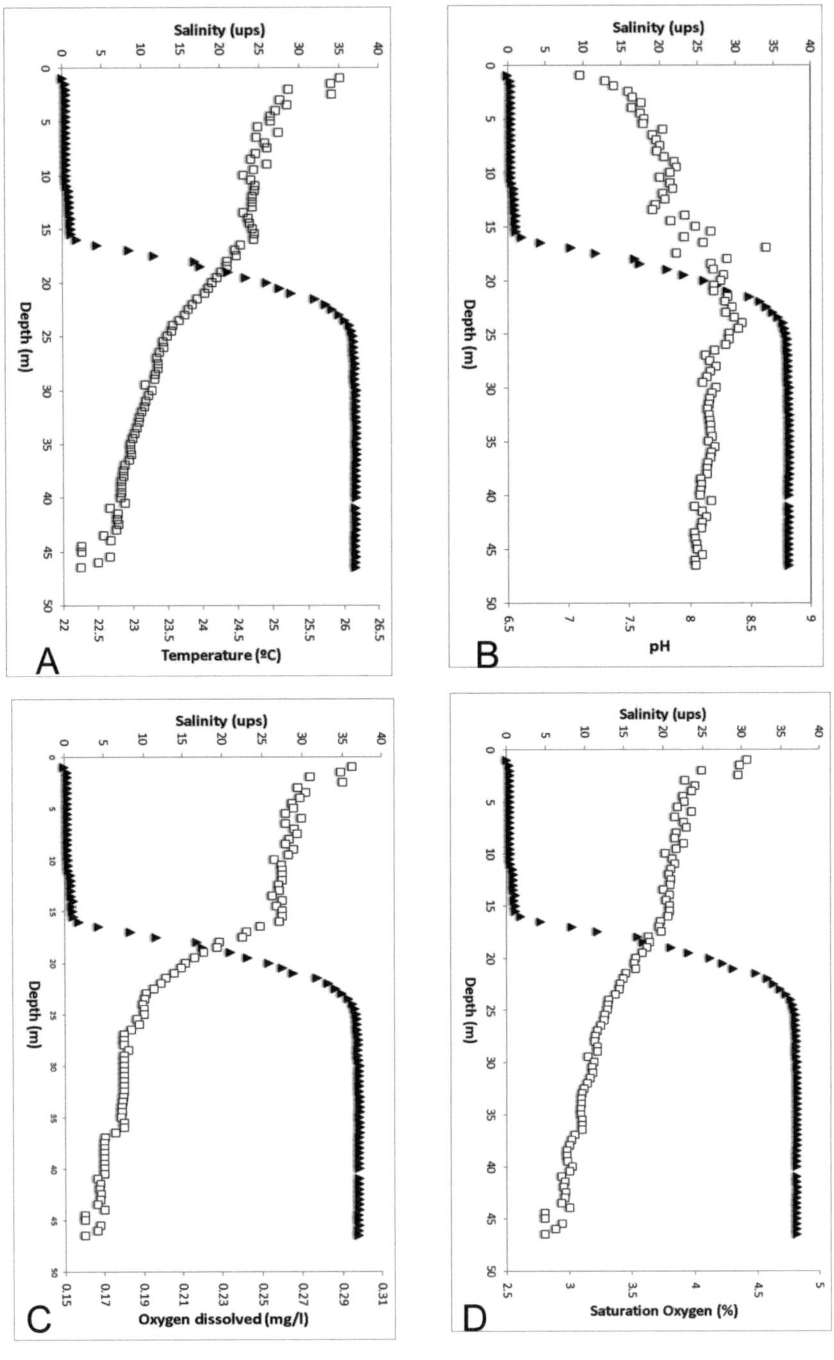

Figure 6.
Profile of Anchialine deep ecosystems. A) Temperature-Salinity, B) pH-Salinity; C) Oxygen Dissolved-Salinity; D) Oxygen concentration-Salinity. The cenote Chemita located in Cozumel Island, have a representative of three water layers i) freshwater 0-16 m; brackish water 16–24m, and marine water 24–60 m.

with more relationships with another Caribbean Islands species live, and normally these species have a microdistribution and marine origin recent. Is evident that this region the species that inhabiting, not share with another region.

In contrast in the Isthmus and Chiapas region the crustaceans have a freshwater origin and their microdistribution are mainly in the locality type for several

Karstic Regions	Species number	Cave number	Specimens recorded	Richness (Cave/sp)	Chao's S_1	Burham & Overton S_2
Tehuantepec Isthmus & Chiapas	16	150	60–90	9.3	16	18.336
Yucatan Peninsula	24	4500	>5000	180	24	25.193
Alta Verapaz	4	18	>1000	4.5	4.5	4.5
Peten	2	5	>500	2.5	2	2
Chiquibul	6	16	>500	2.6	6	6.3

Table 3.
Comparison of richness among karstic regions.

freshwater shrimp and crabs. However according with the geological development from all these regions and this report of species distribution the subterranean corridor existed in this case among Tehuantepec Itsthmus and Chiapas, Alta Verapaz and Chiquibul region for *Macrobrachium* genus, because the all these regions this genus inhabiting the different cave options producing several species that have phylogenetics relationships, at the same situations occurs with Pseudothelphusidae crab family, and of course fishes species. Currently this corridor is not working to gene flow because the caves working like a barrier among populations and the distance are very significative which not happened in the Yucatan Peninsula where the existence of subterranean rivers has been reported.

5. Discussion

The tropical subterranean aquatic biodiversity in Central America (Mexico, Guatemala & Belize) is higher in comparison with those temperate zones in Mexico even, because how has been described previously there are more taxa (**Figures** 7 and **8**) [10, 21, 22, 29]. Although as well had been described, in other continents are described with more detail the taxa numbers by example Europe where the main factor to produce these numbers are the efforts occupied in exploring and registered these taxa [30]. Therefore, the diversity comparison among these regions are too difficult, because depends in first instance of the correct reports and the effort to exploring the areas, all these data are an approximately about that the current status of biodiversity [31], showed the status among this subterranean diversity in the tropics using some cave as examples but is not determinant but in the aquatic habitats the crustaceans are reported as main taxa. However, is evident that the freshwater groups colonised in first instance those habitats close the mountains in this biological subterranean corridor, there are another group that cluster the Yucatan Peninsula Region and due the different origin from the Mountain Systems allow that these species are cosmopolitan distribution [6]. Finally the anchialine species group are totally different and their relationships are more closely with Antilles fauna [32–35]. Even among the regions there are important differences in the diversity not only in the composition of genus or families, too in the number of species and in the specimens registered, by example due that the regions are big land extension is few possible that only one or two specimens were registered, is we check the numbers of index as Chao's or Burman & Overton the diversity estimated increase few because the taxa included in the analysis had a good representation in the area, someone are cosmopolitans in the same region.

Figure 7.
Representative cave crustaceans species in the Mesoamerican region. a) Procambarus *sp.; b)* Typhlopseudothelphusa acanthochela*; c)* Macrobrachium catonium*; d)* Creaseria morleyi *(courtesy by Erick Sosa); e)* Procaris mexicana*; f)* Anchialocaris paulini*; g)* Typhlatya *sp.; h)* Agostocaris zabaletai*; i)* Calliasmata nohochi*; j)* Barbouria cubensis.

By another aspect is the subterranean environments that in first instance was classified in dry, semi-dry, freshwater flooded, and anchialine caves as has been reported in several opportunities [36], but in recently studies has been reported that

Figure 8.
Representative cave fish species in the Mesoamerican region a) Opisternon infernale*; b)* Ogilbia pearsei *(photo courtesy of Juan Carmona; c)* Rhamdia laticauda*; d) fish from Bithynidae family in anchialine ecosystems.*

even in the tropical dry caves there are some different according with the temperature and humidity that produce a major heterogeneity because these features are changing in relation with outside [37, 38], and in semi-dry caves these features have a relation with the oxygen inputs in the subterranean aquatic habitats. In the Anchialine caves the size of freshwater lenses are the main changes to energy entrance and the environmental stability [28]. Their ecological relation in these subterranean systems all in tropical conditions depends of course of their energy support and in this region there are two main ways, the alloctonous way using the biological and hydrological process [36] and for autochthonous way producing by chemolitotrophic procedures the energy using the chemosynthetic bacteria and support by use of methane and dissolved organic carbon [39–41]. In the first option the biological process involve bats that carry several seeds or insect debris, even they self when died; some trogloxene animals that sometimes entrance to these environments and died; but also by there are an important energy sources in cave entrance by the sun effects, where several plants growth and some cave insects go to entrance to feed and back to dark zones. In the second option the chemolitotrophic organisms has been reported on the walls or ceiling as Cueva de Villa Luz in Tabasco and even in some symbiosis with another animals using the electrons from sulphur origin or chemoorganotrophic as the methane decomposition to produce energy as been reported for some crustaceans [42, 43]. This energy source still are working in the different research groups to understand in first instance how is support the life, and the organic matter available could be the main evolutionary forces to different process how as been reported by [15, 21, 44, 45].

6. Conclusions

a. The subterranean diversity in fish and crustaceans species is high in relation with the tropical surrounded environments.

b. The cave decapods is the taxa with major diversity in subterranean Mesoamerican ecosystems

c. The energy to support this diversity had a main source from outside through the carry of organic matter by bats or hydrological r egimens.

d. These ecosystems are strong relationship with outside tropical environments.

Acknowledgements

This manuscript is an outcome of the Biospeleology and Carcinology Lab that has been award for several research funding PRODEP-SEP: Subterranean Biodiversity of Cozumel Island; CONACYT-258494: Molecular systematics of freshwater prawns of genus *Macrobrachium* with abbreviated larval development and their relationship with Guatemala & Belize, UQRoo: Los langostinos del género *Macrobrachium* en México. Also the authors give thanks to those authorities to give permissions to explored and collected animals in the three countries: SEMARNAT, CONANP (Mexico); CONAP (Guatemala); Fisheries and Aquaculture Department (Belize). Special thanks are given to Friends for Conservation, and Belize Audubon Society ONG's from Belize to facilities given to explore Chiquibul and Karstic areas. The authors give thanks to those speleological groups and people that contribute to located the caves in the region as Valerio Sbordoni, Círculo Espeleológico del Mayab, and for their help in the field trip B. Flores, J. Pérez Brückweh, L. Rodriguez, Y. Chable, E. Espadas, X. Rosales, M. Mejia, M. Mejia, J. Cupul, E. Sosa, M. Vazquez, A. Chale, S. Leal, J. M. Tejeda, and to J. Cupul for their help in the data analysis.

Conflict of interest

The authors declare no conflict of interest.

Author details

Luis M. Mejía-Ortíz[1,6]*, Peter Sprouse[2,5], Juan C. Tejeda-Mazariegos[1], Jair Valladarez[3], Oscar Frausto-Martínez[4], Alejandro L. Collantes-Chavez-Costa[5], Guillermo Ruíz-Cancino[1] and German Yáñez[6]

1 Biospeleology and Carcinology Lab, Sustainability Development Division, University of Quintana Roo, Cozumel, Mexico

2 Zara Environmental LLC, Manchaca, TX, USA

3 Belmopan, University of Belize, Belmopan Belize

4 Spatial Observation Lab, Sustainability Development Division, University of Quintana Roo, Cozumel, Mexico

5 Sustainability Development Division, University of Quintana Roo, Cozumel, Mexico

6 Speleological Circle of Mayab, Cozumel, Mexico

*Address all correspondence to: luismejia@uqroo.edu.mx

References

[1] Anonymous. La diversidad biológica de México. Estudio de País. México D. F. Comisión Nacional para el Conocimiento y Uso de la Biodiversidad, 1998: 291p

[2] Mann P. Overview the tectonic history of northern Central America. In: Mann P. (ed). Geology and tectonic development of the Caribbean Plate boundary in northern Central America. The Geology Society of America. 2007: Special Paper 428: 1–19.

[3] López-Ramos E. 1983. Geología de México Tomo III. México D. F. SEP: 1983: 453 p.

[4] Czaplekski NJ, Krejca J, Miller TE. Late quaternary bats from Cebada Cave, Chiquibul Cave System, Belize. Caribbean Journal of Science. 2003: 39(1): 23–33.

[5] Levedeva EV. Karst terrain of the Yucatan Peninsula and it is mountain frame (Mexico, Guatemala & Belize). Geomorfologiya. 2015: 4: 60–79.

[6] Botello A, Alvarez F. Phylogenetic relationships among the freshwater genera of palaemonid shrimp (Crustacea: Decapoda) from Mexico: evidence of multiple invasions?. Latin American Aquatic Journal Research. 2013: 41(4): 773–780.

[7] Arroyave J, Martínez CM, Martínez-Oriol FH, Sosa E, Alter SE. Regional-scale aquifer hydrogeology as a driver of phylogeographic structure in the Neotropical catfish *Rhamdia guatemalensis* (Siluriformes: Heptapteridae) from cenotes of the Yucatan Peninsula, Mexico. Freshwater Biology. 2020: 1–17.

[8] Sbordoni V, Argano R, Zullini A. Biological investigations on the caves of Chiapas (Mexico) and adjacent countries: Introduction. Accademia Nazionale dei Lincei: subterranean Fauna of Mexico Parte II 1973: 1–45.

[9] Sbordoni V, Argano R, Vomero V. Relazione biologica sulle spedizioni Malpaso 1981–82 e 1984. In: Le Spedizioni Speleologiche Malpaso 81 e Malpaso 84 in Chiapas (Messico). Notiziario del Circolo Spelelogico Romano. 1986: 73–88

[10] Reddell JR. A review of the cavernicole fauna of Mexico, Guatemala, and Belize. Bulletin of the Texas Memorial Museum. The University of Texas at Austin.1981: 27: 1–327.

[11] Reddell JR, Veni G. Biology of the Chiquibul Cave System, Belize and Guatemala. Journal of Cave and Karst Studies. 1996: 58(2):131–138.

[12] Mejía-Ortíz LM, Yañez G, López-Mejía M. Fauna of five anchialine caves in Cozumel Island, México. The National Association for Cave Diving Journal. 2006: 39: 11–15.

[13] Mejía-Ortíz LM, Yañez G, López-Mejía M. Echinoderms in an anchialine cave in Mexico. Marine Ecology 28 (Suppl. 1). 2007a: 31–36.

[14] Mejía-Ortíz LM, Yañez G, López-Mejía M, Zarza-González E. Cenotes from Cozumel Island, Quintana Roo, México. Journal of Cave and Karst Studies. 2007b 68(2) 250–255.

[15] Mejía-Ortíz LM. Adaptaciones de los crustáceos a la vida subterránea. In: Rodríguez-Almaráz G, Alvarez F. (Editors). Crustáceos de México: Estado actual de su conocimiento. Monterrey, Universidad de Nuevo León-Instituto de Biología UNAM 2009: 419–458 pp

[16] Yañez-Mendoza G, Zarza-González E, Mejía-Ortíz LM. 2008. Sistemas anquihalinos. In: Mejía-Ortíz LM. (Editor) 2008. Biodiversidad acuática de la Isla de Cozumel. México D. F. Plaza y Valdes/

Universidad de Quintana Roo, 2008: 49–72 pp

[17] Gambari S. Le cavita esplorate in Messico (1986-87) rilievi e descrizioni morfologiche. In: Le Spedizioni Speleologiche Malpaso 86 e Rancho Nuevo 87 in Chiapas (Messico). Notiziario del Circolo Speleolgico Romano, Nouva Serie. 1987: 2: 87–134.

[18] Bauer-Gottwein P, Gondwe BRN, Charvet G, Marin LE, Rebolledo-Vieyra M, Merediz-Alonso G. Review The Yucatan Peninsula karst aquifer Mexico. Hydrogeology Journal 2011: 19(3): 507–524.

[19] Rampini M, Di Russo C. Report Bioespeleologico su algune groote del Guatemala. In: Spedizioni in Messico (Chiapas) e in Guatemala dal 1996 al 2001. Notiziario del Circolo Speleologico Romano. 2004. Nouva Serie 16-19: 93–100.

[20] Tejeda-Mazariegos JC, Mejía-Ortíz LM, López-Mejía M, Crandall KA, Pérez-Losada M, Frausto-Martínez' O. Freshwater crustaceans decapods an important resource of Guatemala. In: Sajal R. (Editor) Biological Resources of Water Serbia. InTech Publisher. 2018. 169–179 pp.

[21] Mejía-Ortíz LM, López-Mejía M, Pakes J, Hartnoll R, Zarza-González E. Morphological adaptations to anchialine species of five shrimp species (*Barbouria yanezi, Agostocaris bozanici, Procaris mexicana Calliasmata nohochi and Typhlatya pearsei*). Crustaceana. 2013a: 86(5): 578–593.

[22] Mejía-Ortíz LM, López-Mejía M, Sprouse P. Distribución de los crustáceos estigobiontes de México. Mundos Subterráneos UMAE. 2013b: 25: 20–32.

[23] Mejía-Ortíz LM, López-Mejía M, Tejeda-Mazariegos JC, Frausto-Martínez O, Crandall KA, Pérez-Losada M, Valladarez JG. Los camarones de agua dulce de la subfamilia Palaemoninae en la Península de Yucatán, (México, Guatemala y Belice). Teoría y Praxis. 2018: 25: 115–130

[24] Mejía-Ortíz LM, Cupul-Pool JE, López-Mejía M, Baez-Meléndres AG, Tejeda-Mazariegos JC, Valladarez JG, Crandall KA, Pérez-Losada M, Frausto-Martínez O. The habitat types of freshwater prawns (Palaemonidae: *Macrobrachium*) with abbreviated larval development in Mesoamerica (Mexico, Guatemala and Belize). In: Diarte-Plata G, Escamilla-Montes R. (Editors) Crustacea. Croacia. Intech Press. 2020. 77–87 pp.

[25] Schneider K, Culver DC. Estimating subterranean species richness using intensive sampling and rarefaction curves in a high density cave region in West Virginia. Journal of Cave and Karst Studies 2004. 66(2): 39–45.

[26] Chao A. Non-parametric estimation of the number of classes in a population: Scandinavian Journal of Statistics. 1984. 11: 265–270.

[27] Burnham KP, Overton WS. Estimation of the size of a closed populations when capture probabilities vary among indivisuals. Biometrika 1978. 65: 623–633.

[28] Mejía-Ortíz, LM, Arriaga-Velez MD, Yánez-Mendoza, G. Environmental heterogeneity from anchialine caves: Biodiversity and communities composition. Special Issue Research on Karst Ecosystem. In prep

[29] Mejia-Ortiz LM. Cave crustacean decapods from Mexico. In: Lachance N. (Editor). The Zoological Guide to Crustacea. New York. Nova Science Publisher. 2019: 1–69

[30] Deharveng L, Gibert J, Culver DC. Biodiversity in Europe. In: Withe WB,

Culver DC, Pipan T. (Editors). Encyclopedia of Caves. Oxford. Third Edition. Academic Press. 2019: 136–145.

[31] Deharverg L, Bedos A. Biodiversity in the tropics. In: Withe WB, Culver DC, Pipan T. (Editors). Encyclopedia of Caves. Oxford. Third Edition. Academic Press. 2019: 146–162.

[32] Alvarez F, Iliffe TM, Benitez S, Brankovits D, Villalobos JL. New records of anchialine fauna from the Yucatan Peninsula Mexico. Checklist. 2015; 11(1) 1505: 1–10

[33] Mejía-Ortíz LM, Yañez G, López Mejía M. Anchialocarididae, new family of anchialine decapod and a new species of genus *Agostocaris* from Cozumel Island, México. Crustaceana. 2017: 90 (4): 381–398.

[34] Ditter RE, Mejía-Ortíz LM, Bracken-Grissom HD. Anchialine Adjustments: an updated phylogeny and classification for the family Barbouriidae Christoffersen, 1987 (Decapoda: Caridea). Journal of Crustacean Biology. 2020: 40(4): 401–411.

[35] Sket B. Anchihaline (Anchialine) caves and fauna. In: Withe WB, Culver DC, Pipan T. (Editors). Encyclopedia of Caves. Oxford. Third Edition. Academic Press. 2019: 56–64.

[36] Culver D, Pipan T. The biology of caves and another subterranean habitats. Oxford, Oxford University Press. 2010: 254p

[37] Mejía-Ortíz LM, Pipan T, Culver DC, Sprouse P.. The blurred line between photic and aphotic environments: a large Mexican cave with almost no dark zone. International Journal of Speleology. 2018: 47(1): 1–12.

[38] Mejía-Ortíz LM, Christman MC, Pipan T, Culver DC. What's temperature in the tropical caves?. PlosOne. 2020: 15(12): 1–21

[39] Kumaresan D, Hillebrand-Voiculescu AM, Wischer D, Stephensen J, Chen Y, Murrell JC. Microbioal life in unusual cave ecosystems sustained by chemosynthesis primary production. In: Summers Engel A (Ed). Microbial life of cave systems. Life in extreme environments. Berlin, De Gruyter, 2015: Berlin. 215-229p.

[40] Brankovits D, Pohlman JW, Niemann H, Leigh MB, Leewis MC, Becker KW, Iliffe TM, Alvarez F, Lehmann MF, Phillips B. Methane- and dissolved organic carbon-fueled microbial loop supports a tropical subterranean estuary ecosystem. Nature Comunications, 2017. 8:1835: 1–12.

[41] Chávez-Solís EM, Solís C, Simoes N, Mascaró M. Distribution patterns, carbon sources and niche partitioning in cave shrimps (Atyidae: Typhlatya). Scientific Reports Nature, 2020. 10: 12812: 1–16.

[42] Pakes MJ, Weiss AK, Mejía-Ortíz LM. Arthropods host intracellular chemosynthetic symbionts, too: Cave study reveals an unusual form of symbiosis. Journal of Crustacean Biology. 2014: 34(3): 334–341.

[43] Pakes MJ, Mejía-Ortíz LM. Chemosynthetic ectosymbiosis reported in the predatory anchialine cave endemic, *Xibalbanus tulumensis* (Yager, 1987) (Remipedia). Crustaceana. 2014: 87(14): 1657–1667.

[44] Mejía-Ortíz LM. Adaptations to cave life in decapods from Oaxaca. Austin. Association for Mexican Cave Studies Bulletin 15. 2005: 170 pp.

[45] Mejia-Ortíz LM, López-Mejía M, Bribiesca-Contreras G, Solís-Marín FA, Yañez G. La faune anchialine de l'ile de Cozumel In: Thomas C., Les grottes du Yucatan: Ile de Cozumel. Montreuil. Editions Xibalba. 2013c: 140–155 pp

Chapter 4

A Natural History of Floating *Sargassum* Species (Sargasso) from Mexico

José Luis Godínez-Ortega, Juan V. Cuatlán-Cortés, Juan M. López-Bautista and Brigitta I. van Tussenbroek

Abstract

For at least several centuries, sargasso has inhabited the Atlantic Ocean, and there are historical records of these algae reaching the Mexican Veracruz State in the Gulf of Mexico. Blooming of sargasso in the southern tropical Atlantic is a current a global problem from Africa to the Greater Caribbean. Since 2015, exceptionally large quantities of sargasso have been arriving intermittently on the Mexican Caribbean coast, affecting coastal ecosystems and tourist beaches. Sargasso includes two holo-pelagic species, *Sargassum natans* and *S. fluitans*, with several varieties. There are no records of sexual reproduction in these species, and the algae are thought to spread exclusively by clonal reproduction by fragmentation. Although sargasso seaweeds have grown in the Sargasso Sea for centuries; they have not been well studied. This chapter deals with historical aspects of these algae, their taxonomic and morphological characteristics, distribution, ecology, and practical uses. Sargasso blooms in the central Atlantic started in 2011. In later years, the bloom developed to extend from West Africa, Brazil, and the Great Caribbean, including West-Indies, Mexico, and the Gulf of Mexico. The pelagic sargasso is a global phenomenon that must be understood by integrating natural history, modern biology, social and economic aspects.

Keywords: pelagic *Sargassum*, bloom, Sargasso Sea, Great Atlantic Sargasso Belt, uses, historical background, algal biology

1. Introduction

The Mexican beaches of the western Atlantic Sea, particularly those of the Mexican Caribbean in the state of Quintana Roo, are among the most visited by national and international tourists due to their beautiful turquoise waters and white calcareous sand. These beaches have recently been infested by huge drifting rafts of macroalgae the genus *Sargassum* (Phaeophyceae); i.e. *Sargassum fluitans* and *S. natans*. In the open ocean these rafts are refuge, substrate or nursery for many marine flora and fauna. However, the quantities reaching the Mexican beaches intermittingly in recent year have been massive, covering vast coastal areas, causing severe problems to marine ecosystems and tourism [1–4]. The massive influxes and beaching of these seaweeds have become a persistent phenomenon, which has led to a series of investigations in Mexico and other parts of the world, that aim to understand this phenomenon and to develop adaptation strategies to mitigate its damage to the ecosystems and economy.

This problem has prompted us to ask ourselves various questions of a natural history nature: How long have these seaweed rafts been around? Have similar bloom events occurred in the past? What is the origin of the recent blooms? These are essential questions to understand the recent infestation and their answers may aid to further research into mitigation of its effects in the short, medium, and long term. We will do so from an historical and biological perspective, starting with the algal names, which today seems trivial but has a fascinating history beginning in the 15th century. We also recognize that it is crucial to know their correct scientific name as the first step to study their biology. Many herbarium specimens have been collected in the past and are awaiting to be investigated in terms of their macroscopic morphology, internal anatomy, molecular biology, reproduction, ecology, and geographical distribution. We will also provide information on past and present uses of sargasso.

1.1 Fossil and historical records of sargasso

The fossil record of sargasso species assemblages goes possibly back to the Tethys Sea. Jerzmańska and Kotlarczyk [5] described numerous brown algae (Phaeophyceae), together with fish skeletons from the Oligocene (33.9–23.03 mya) in the Polish Carpathians [5]. These authors pointed out that in this fossil record some algae were observed with air bladders inserted on branches, but without holdfasts or attachment structures, similar to modern "pelagic forms of the genus *Sargassum*" [5]. The authors hypothesize that the fish genera in this thanatocoenosis fossil assemblage, belonging to the upper part of the upper bathypelagic horizon (seabed at 1000 to 4000 m deep), might be formed at the bottom of a sargasso sea, with the presence of pelagic sargasso, bathypelagic and pelagic fishes comparable to those of recent marine biocenoses of the Sargasso Sea. They called this Oligocene assemblage "quasi-Sargasso". Since there are no fossil records from the North Atlantic, the hypothesis presented in this work is that the "quasi-Sargasso" assemblage had its origin in the Tethys Sea and subsequently migrated towards the Atlantic Ocean before its original habitat was destroyed by alpine folding at the end of the Miocene [5]. Despite early fossil records of sargasso-like algae, extensive phylogenetic analysis indicates that the diversification of the genus *Sargassum* was likely relatively recent, not before the late Pliocene (5.3–2.6 mya), with its origin in the central Indo-Pacific, and diversification into the Atlantic 0.2 to 0.4 million years ago [6]. Following this hypothesis, the fossils found in the Oligocene Carpathians' beds corroborate the existence of an ancestor of the genus *Sargassum* 36 million years ago, with diversification in the late Pliocene. From this, we can tentatively conclude that the pelagic species were probably present in the Atlantic long before the diversification of benthic species, which will need verification with further future studies. Børgesen [7] mentioned that the pelagic *Sargassum* species were of benthic origin, but Parr [8] pointed out that the lack of benthic species with comparable morphology to that of the pelagic ones in the western Atlantic, casts doubt on Børgesen's hypothesis. The oldest written observation on sargasso are from Columbus; thus, there is a great time gap between fossil records and actual sightings.

Pérez-Rubín Feigl [9] cited sightings of sargasso by inhabitants of the American continent before Christopher Columbus' voyage in 1492 and up to 1792 in his detailed study on "Las algas y los antiguos navegantes españoles (1492-1792)" ["Algae and ancient Spanish navigators (1492-1792)"]. In Mexico, there is a record of sargasso by the ancient Mayans. They call sargasso "U tail kaknab", which means "is thrown by the lady of the sea". Possibly, this is the reason for a present-day confusion in the Yucatan Peninsula about the term "sargazo" which is the local name for all types of plant material found on the beach and not only sargasso or other *Sargassum* spp. [10]. But, it is very certain that they had seen sargasso among the other beach-cast specimen. Christobal Columbus, in his "Relaciones y Cartas," from his first trip, mentioned the following: "en amaneciendo hallaron tanta yerba que parecía ser la mar cuajada de ella, y venía del Oeste" (Viernes 21 septiembre 1492) ["at dawn, they found so much grass that it seemed to be the sea curdled with it, and it came from the West" (Friday, September 21, 1492)] [11].

Sargasso is a brown pelagic alga kept afloat by their small air-filled bladders (pneumatophores), that can form large entangled assemblages. The Portuguese ships were often entrapped in the algal masses due to the lack of wind, giving crews plenty of time to explore it. As these men came from a country where vines abound, the air bladder assemblages of the seaweeds seemed bunches of grapes to them, of a variety called "salgazo". "Sargacinha" means grape, which comes from "sarga" (variety of grape) [12], and finally, it was derived in "sargaço" or "sargaçao" [13]. This may be how this alga acquired its name. In another letter dated October 3, 1492, Columbus commented on these algae and mentioned: "Aparecieron parcelas, yerba mucha, alguna muy vieja y otra muy fresca, y traía como fruta;" ["Plots appeared, a lot of herbs, some very old and some very fresh, and they brought like fruit," [11]. It is probable that at this time, the Sargasso Sea was named.

The name sargasso may have yet another origin than the one described above. It is known that in Portugal, from the Middle Ages to the 20th century, the harvesting of algae was an economically and socially significant activity. In a letter dated March 9, 1308, D. Dinis orders that the "argaço" that came from the sea, belonged to the residents of the place, which they prepared and dried to fertilize the fields [14, 15]. Thus, the sailors on Columbus ships may have called them"argaço" or "salgazo". In this case, the evolutionary line of the word alga> algaço> argaço and sargaço, is thought to derive from argaço under the influence of another word, probably salt [16, 17].

Between 1526 and 1590, Gonzalo Fernández de Oviedo (1478–1557) and José de Acosta (1540–1600) disseminated the natural sciences of America, and their texts had a worldwide distribution at that time. After the Columbian period, authors such as Juan López de Velasco (1530–1598), Bartolomé de Las Casas (1474 or 1484–1566), Gonzalo Fernández de Oviedo (1478–1557), José de Acosta (1540–1600), Pedro Martir de Angleria (1457–1526) in their Decades (1515) and Alexander von Humboldt (1769–1859) mentioned the "mar de hierbas" [sea of herbs] or *Mare Herbidum* [9].

The description by Fernández de Oviedo in 1535 [18] stands out because he mentioned "the great grassland" and named the algae on the surface of the sea "salgazos". Acosta [19] wrote the following: "En la muy profunda y larga mar de la muy nombrada, y no menos temida Vuelta del Sargazo (que así se llama de los navegantes de las Indias de diez y ocho hasta treinta y cuatro grados de la línea equinoccial de la parte del Norte) aparece la mar llena de esta yerba, llamada sargazo. Es de un palmo: los ramillos delgados y sin raíz: véase toda el agua cubierta de esta yerba en montones pegada y liada una con otra y especulando bien le ve venir del profundo de la mar tan liada, y envuelta, que parece cada montón una grande mata." ["In the very deep and long sea of the much named, and no less feared *Vuelta del Sargazo* (which is the name of the navigators of the Indies from eighteen to thirty-four degrees of the equinoctial line of the northern hemisphere) appears the sea full of this herb, called sargasso. It is of a hand span (20 cm): the thin and detached branchlets: see all the water covered with this herb in heaps stuck together and bundled with each other and, speculating well, you see it coming from the deep sea so bundled, and messy, that each heap seems one big clump."].

Alexander von Humboldt (1769–1859) was the first scientist to study the Sargasso Sea in more detail, situating it in the eastern and American region, the latter with the highest concentration, placing it between Bermuda and the Bahamas. The area was well known to sailors: "An ancient tradition, which has been preserved among the pilots of Galicia, says that this large bank of "fucus" marks the middle of the route they take through the “Golfo de las Yeguas” [“Gulf of Herbs“] the ships return to Spain from Cartagena de Indias, Veracruz or Havana, which are favored by the current of the Gulf Stream” [20]. Humboldt interpreted this Sargasso Sea as a community association, constituted by the algal species and an animal community [9, 20]. Hipólito Ruiz López (1754–1816), a Spanish botanist, published his study on sargasso and pointed out the distribution in the Atlantic Ocean between 22° to 38°N. In this study, he illustrated *Fucus natans* and confused its epibionts (hydrozoid) with anthers and pistils from vascular plants and the pneumatocysts with seeds (**Figure 1**) [21]. Sargasso was mentioned in 1799 by José de Viera y Clavijo

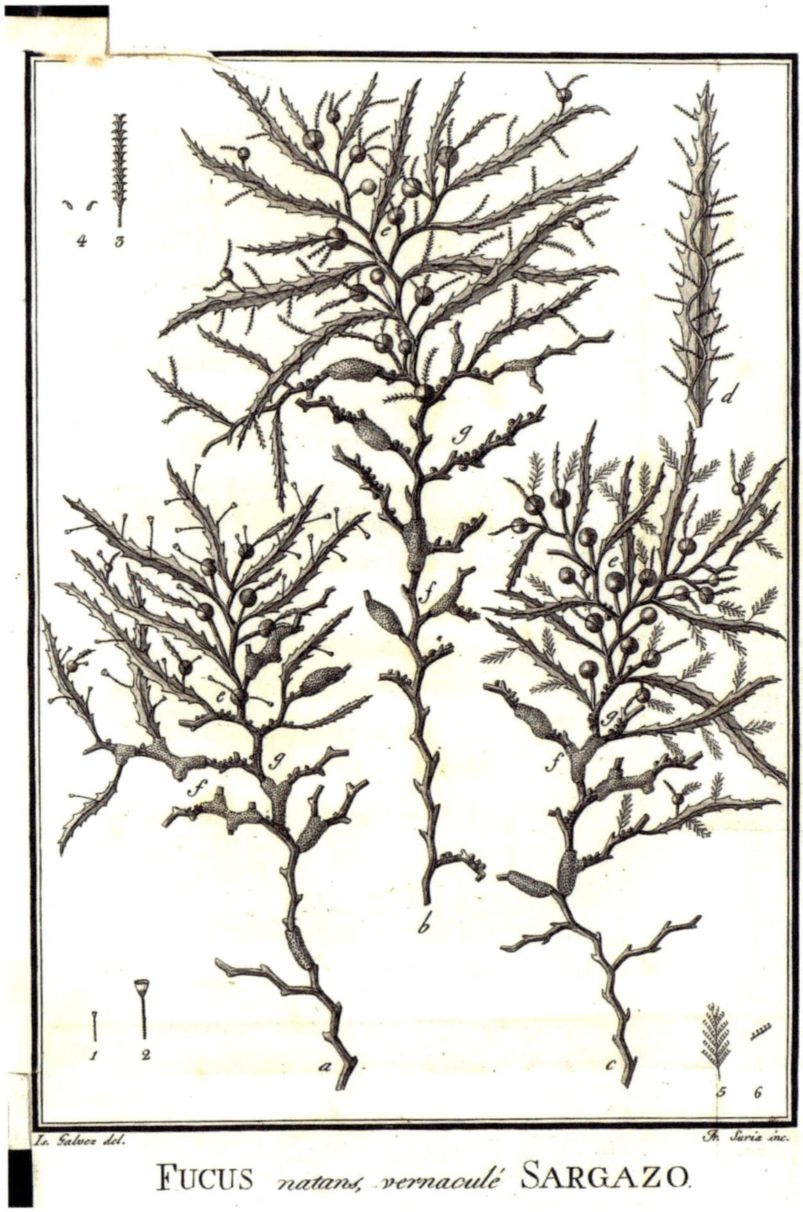

Figure 1.
Fucus natans L. plate taken from the taxonomic description of Hipólito Ruiz (1798) [21]. Digital library of the Royal Botanical Garden (Madrid, Spain).

(1731–1813), who presented morphological descriptions and characteristics of its habitat. Viera and Clavijo's works were published in the 19th century [22, 23].

In the 18th century, Martín de Sessé and José Mariano Mociño, members of the Royal Botanical Expedition of New Spain, recognized algae and particularly sargasso. In the "Catálogo de los animales y plantas que han reconocido y determinado según el sistema de Linneo los facultativos de mi expedición D. José Mociño y D. José Maldonado" [24]. This document was found in the Archivo General y Biblioteca del Ministerio de Asuntos Exteriores from Madrid as a manuscript: "Viaje a la

Figure 2.
Sargassum bacciferum (Turner) C. Agardh (S. natans) collected by F.M. Liebamann (1841–1843) from Campeche Bank [C-A-99772] [25].

Species	State	Locality	Region	Coll.	Date	Morpho-types	Herb.	Ref.
Sargassum fluitans	Tamaulipas	Boca del Río Bravo, Escollera N del Puerto El Mezquital	Shore	A. Schott;S. Martínez Lozano & J.M. López Bautista	1853; 1991		F	[26–28]
	Veracruz	Tuxpan, Tampamachoco Lagoon, Tamiahua Lagoon, Hermosa Beach, Muñecos Beach	Shore	S. de la Campa; D. González Nieto	1965; Sep. 2011		UAMIZ	[29, 30]
	Veracruz	Enmedio Reef	Ocean	L. Huerta	Apr. 1965; 1971–1974			[31, 32]
	Veracruz	Anegada de Afuera Reef	Ocean	J. L Godínez	March 18, 2014	III	2458, 2459, 2460, 2461 (MEXU)	This study
	Campeche	Arcas Key, Triángulo Oeste	Ocean	L. Huerta	1987			[33]
	Yucatan	Progreso	Shore	L. Huerta	1987			[33]
	Yucatán	Perez Island, Alacrán Reef, Arenas Key, Alacrán Reef	Ocean	L. Huerta; J.T. Conover & W.D. Perkins	Mar. 1955; 1959; 1960–1961; 1987			[33–36]
	Quintana Roo	Punta Hualapich, Puerto Morelos, Puerto Xcalak, Tulum, Mahahual, Xahuayxol, Paamul, Cancun	Shore	L. Huerta; L. Collado Vides et al.; K. Dreckmann; D. González Nieto & A. Sentíes; Aguilar Rosas bros.	1971–1974; 1986; 1987; 1990; 1993–1995; Feb. 1994; April. 1995; Mar. 2012; Sep. 2016; Aug.-Dec. 2018; Apr. 2018-Mar. 2019; Apr.-Jun. 2019; May 2019	III	UAMIZ; FCME-Sisal; ECOSUR	[33, 37–44]

Species	State	Locality	Region	Coll.	Date	Morpho-types	Herb.	Ref.
		Puerto Morelos	Shore	J. Chalé et al.; F. de Lorens	Feb. 18–19, 2012; Jun. 20, 2019	III	2042, 3532 (MEXU)	This study
		Cancún	Shore	Leg. S. Figueroa	November 25, 2015	III	2385, 2447, 2448	This study
		Tulúm	Shore	M. Pace	June 29, 2019	III	3551 (MEXU)	This study
	Quintana Roo	Cancun Island, Mujeres Island, Cozumel Island	Ocean	D. González Nieto & L.E. Mateo-Cid	1987; May 2013; May 2019	III	UAMIZ	[29, 33, 38]
	Gulf of Mexico		Ocean		1939; Apr. 1956			[8, 45, 46]
Sargassum natans	Tamaulipas	Boca del Río Bravo	Shore	A. Schott; S. Martínez Lozano & J.M. López Bautista	1853; 1991		F	[26, 28]
	Veracruz	Hotel Pensiones Beach, Veracruz Harbor, Hermosa Beach	Shore	F.M. Liebmann; D. González Nieto	1841–1843; Jun. 1964; Sep. 2011		C; UAMIZ	[29, 30]
		Boca del Río	Shore	Munguía et al.	Apr. 8, 2011	I	1045 (MEXU)	This study
		Ingeniero Reef	Shore	G. Almeida et al.; J.C. Blanco Camarillo et al.	Apr. 13 2012; May 17, 2013	I, VIII	3603, 3604 (MEXU)	This study
		El Morro de La Mancha	Shore	I. Alonso et al.	Oct. 22, 2011; Apr. 26 2014	I	2456, 3605, 3606 (MEXU)	This study
		Punta Montepio	Shore	Workshop Level 3	Feb. 13, 2014	I	2117, 2121 (MEXU)	This study
		Enmedio Reef	Ocean	R.F. Hohenacker	1888; 1957–1959; 1964; 1971–1974; 1992		BM	[30–32, 47–49]

Species	State	Locality	Region	Coll.	Date	Morpho-types	Herb.	Ref.
	Campeche	Arcas Key, Triángulo Oeste, Campeche Bank	Ocean	F.M. Liebmann	1841–1843; 1987		C	[25, 33]
	Yucatan	Progreso	Shore	L. Huerta	1987			[33]
		Arenas Key, Alacrán Reef	Ocean	L. Huerta; J.T. Conover & W.D. Perkins	1987; July 1960			[33, 36]
	Quintana Roo	Xcalak, Puerto Morelos, Mahahual, Xahuayxol, Punta Estrella, Akumal, Xcacel, Tulum	Shore	L. Huerta & A.M. Garza-Barrientos; L. Collado Vides et al.; R.E. Rodríguez Martínez; M.A. Mendoza Becerril; M. García Sánchez; Aguilar Rosas bros.	Apr.-Jun. 2019; May 2020; 1971–1972; 1986; 1987; Apr. 1995; Sep. 2016; Aug.-Dec. 2018; Apr. 2018-Mar. 2019	I, VIII	FCME-Sisal; ECOSUR	[33, 37–39, 41, 43, 44]
		Tulum	Shore	M. Pace	July 27, 2019	I	3595 (MEXU)	This study
		Chinchorro Bank, Mujeres Island, Conty Island, Cancun Island, Cozumel Island, Agua Azul	Ocean	L. Huerta; R.E. Rodríguez Martínez	1987; Aug. 2018; May 2019	I, VIII		[33, 38]
	Atlantic Coast of Mexico		Ocean	H.H. Hildebrand	1954–1957; 1960; 1962			[45, 46, 50]

Table 1.
Records of floating Sargassum spp. in Mexico.

costa Noroeste de la América Septentrional, por don Juan Francisco de la Bodega y Quadra ..." from 1792. In this list we found *Fucus natans*, currently *Sargassum natans*, it is clear that it is sargasso, very possibly collected in the Gulf of Mexico, perhaps to drift; unfortunately, the exsiccata is not found in the Herbarium of the Royal Botanical Garden of Madrid.

The first published records we have in Mexico are from Veracruz and the Campeche Bank collected by F. Liebmann in the period 1841–1843 (**Figure 2**). Liebmann was a Danish naturalist and his collections were studied by the Swedish Phycologist C. Agardh [25]. Sargasso sightings have been reported from very ancient times to the present day. The data obtained from natural history collections in Mexico begin in 1841 and last until 2019. These records that are found in institutional herbaria and that have also been published are presented in **Table 1**. Records were located in the 5 coastal states of Mexico (Tamaulipas, Veracruz, Campeche, Yucatán and Quintana Roo), both coastal sargasso and the oceanic zone. The first record we have is from Veracruz and the Campeche Bank collected by F. Liebmann in the period 1841–1843 (**Figure 2**). Liebmann was a Danish naturalist and his collections were studied by the Swedish Phycologist C. Agardh. In the 50's and up to the 80's (20th century), collections of *Sargassum* spp. by Laura Huerta and María Ana Garza Barrientos. In the 90s, the studies of K. Dreckmann, Ligia Collado and A. Sentíes stand out. Studies continue in the 21st century, but mainly due to the massive arrivals that began in 2011 until 2019.

2. Biology of sargasso

2.1 Methodology of the study of herbarium specimen

The material deposited in the National Herbarium (MEXU) of the Institute of Biology, National Autonomous University of Mexico (UNAM) (**Table 1**) was reviewed. Herbarium samples were hydrated for 24 h with seawater and a liquid soap solution (5%). Once hydrated, they were fixed in a 4% formalin solution neutralized in seawater. For morpho-anatomical observations, 15 μm-thick sections were made with a microtome (Reichart Jung 820) and a Mectron cryostat with disposable blades. Sections were mounted in a 70:30 Karo® corn syrup/water solution with a trace of phenol to prevent fungal growth [51]. A Zeiss light microscope, model 1206 S09432, equipped with a Canon PowerShot G6 digital camera, was used to determine the alga's morphological characteristics. For cell measurements AxioVision software (SE64, Rel. 4.9.1 Carl Zeiss) was used. The species descriptions below are based on this material.

2.2 Classification and species descriptions

The study of the genus *Sargassum* began with the work of C. Agardh in 1820 [52]. However, this name was born from the *Fucus* of Linnaeus [53]. The basionym *Fucus natans* was mainly applied to sargasso from the Atlantic Ocean or "Common Gulfweed," while *S. fluitans* was named "Broad-toothed Gulfweed."

The taxonomic and nomenclatural history of *S. fluitans* and *S. natans* is explained extensively by Silva et al. [54]. The genus *Sargassum* from the order Fucales is a diversified genus with almost 361 species [55]. Identification of the species may be problematic due to their polymorphic character and phenotypic plasticity [56]. Traditionally, classification was based on the blade morphology, the margins of the blades, blade midrib, the pneumatophores (air bladders), the branching, degree of branching, and morphology of the receptacles (reproductive organs) [57].

The genus *Sargassum* has received increasing attention since 1985 [57]. At first, four subgenera were recognized: *Sargassum*, *Arthrophycus*, *Bactrophycus*, and *Phyllotrichia*. The subgenus *Sargassum* is the most diverse one, and it is divided into three sections: Zygocarpicae, Malacocarpicae, and Acanthocarpicae [57]. Recently, the traditional classification has been put to the test using nuclear, chloroplast, and mitochondrial markers. One of the resulting outcomes was polyphyly of the section Acantocarpicae, which was subsequently considered as a synonym of the section *Sargassum*. Also, based on molecular analysis, *S. natans* was placed in section *Sargassum*, because its receptacle morphology cannot be verified [57] since until date only its vegetative form is known. At present, the genus has only two recognized subgenera: i.e., *Sargassum* and *Bactrophycus*, with nine sections found in temperate and tropical latitudes [58].

In Mexico, both pelagic species, *S. fluitans* and *S. natans* have been recorded [50, 59]. According to Parr [8], there are three morphotypes or ecotypes for the Mexican region.

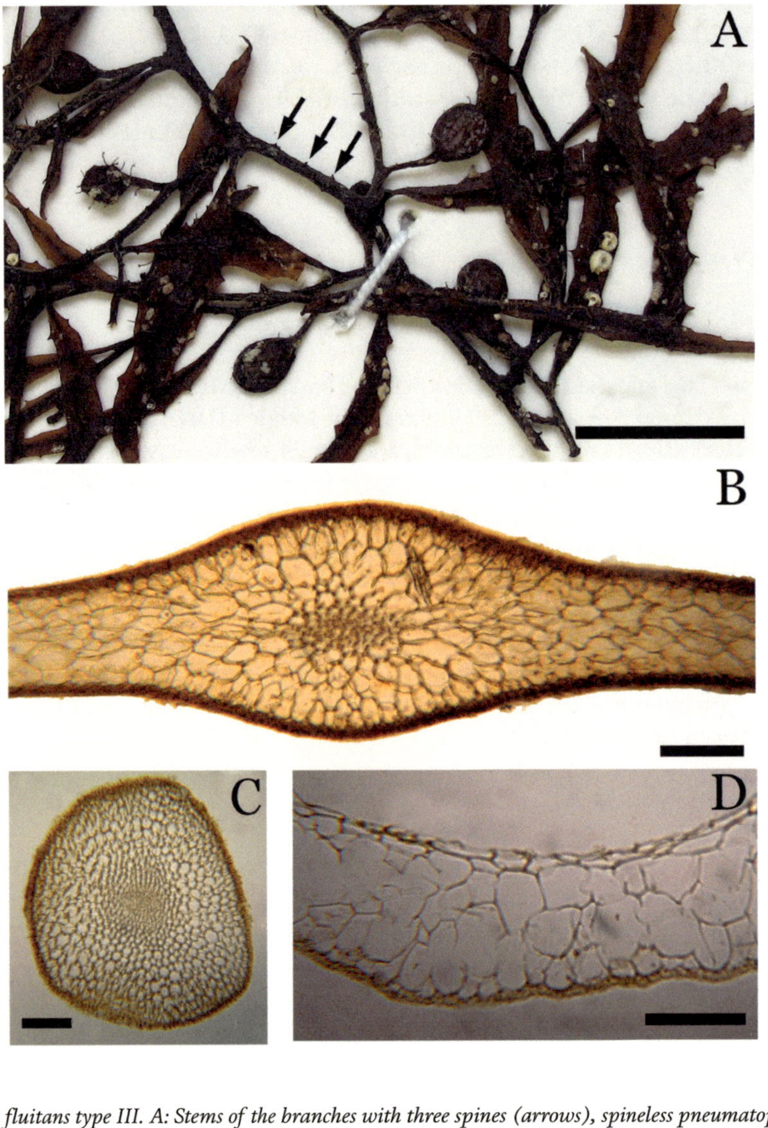

Figure 3.
Sargassum fluitans type III. A: Stems of the branches with three spines (arrows), spineless pneumatophores with many more or less wide blades (NI-MEXU 2459); scale 2 cm. B: Cross section of the leaf shows the midrib and a cell layer in the cortex. C: Cross section of the stem. D: Cross section of the pneumatophore formed from a thick layer of cells. Scale bars: 200 μm.

It is essential to point out that Parr [8] did not follow the International Code of Botanical Nomenclature [60] while designating its types, so his designations are considered an artificial classification. However, due to the massive sargasso invasions on the western Atlantic Ocean, Parr's classification of the morphotypes is useful.

Classification: Phylum Ochrophyta; Class Phaeophyceae; Subclass Fucophycidae; Order Fucales; Family Sargassaceae; Genus *Sargassum*; Subgenus *Sargassum*; Section *Sargassum*.

Descriptions from Mexican material:

Sargassum fluitans (Børgesen) Børgesen, 1914b: 66, footnote. Type locality: Sargasso Sea [7]. **Figures 3** and **4**.

Homotypic synonym:

Sargassum hystrix J. Agardh var. *fluitans* Børgesen, 1914a: 11 [61].

Figure 4.
Sargassum fluitans type III (Puerto Morelos). Scale bar: 1 cm.

Pelagic specimens (up to 1 m; usually 20–30 cm long) of yellowish-brown color are forming rafts of variable sizes. Cylindrical axis 0.4–1.8 mm thick, smooth or with few spines near the apex, with a well-formed midrib in the center surrounded by rounded cells and a single layer cortex of cells (**Figure 3A,C**). Blades with short pedicels of 1–5 cm long by 1–4 mm wide, 423 μm thick in the distinctive midrib; blades firm, lanceolate in shape and pointed apices, serrate margins with broadly flattened teeth at the base; in cross-section, a well-formed midrib is observed in the center surrounded by polygonal medullary cells protected with a cortex of a layer of quadrangular cells (**Figure 3 A,B**). Cryptostomatas (sterile cavities with hairs) absent. Pneumatophores at the base of the blades 2–5 mm in diameter, with a pedicel of 3–4 mm without wings; in cross section with two cell layers, the outer cortex with smaller cells (**Figure 3 A, D**). Unknown receptacles.

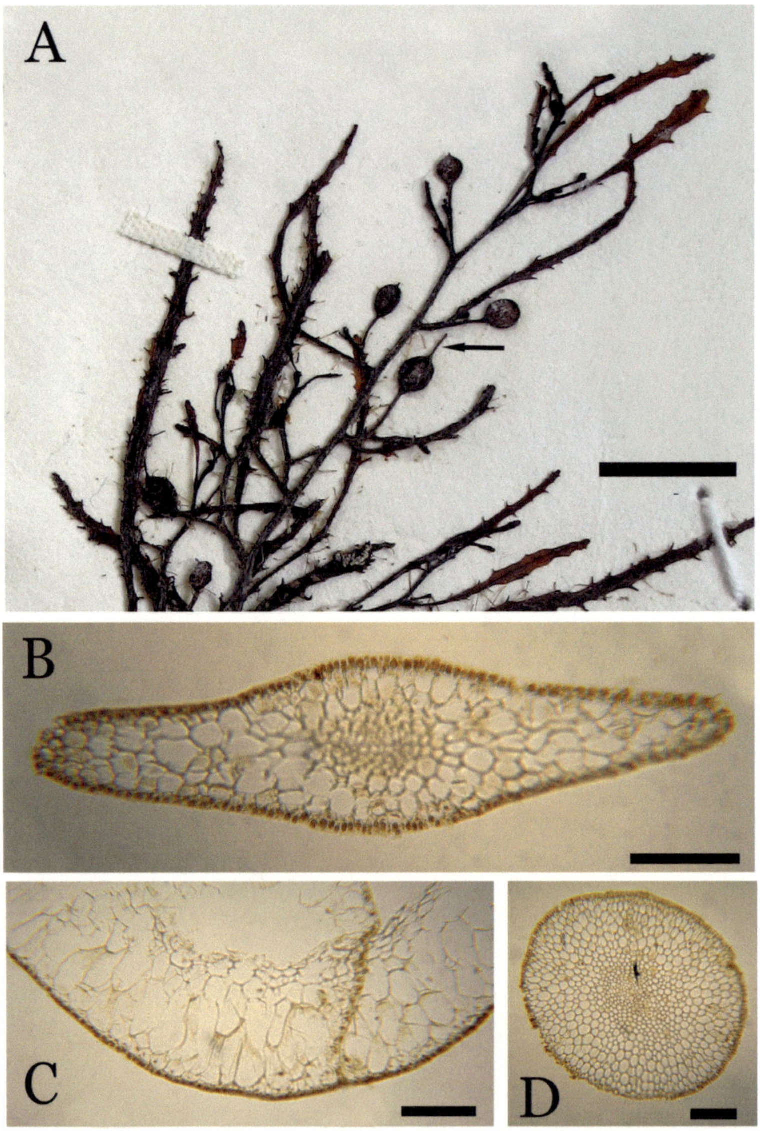

Figure 5.
Sargassum natans type I. A: Smooth branches with distal spine on pneumatophores (arrow) and narrow blades (NI-MEXU 2603). B: Cross section of the blade shows the midrib and a cell layer in the cortex. C: Cross section of the pneumatophore formed by a thick layer of cells. D: Cross section of the stem. Scale bars: 200 μm.

Sargassum fluitans Type III [8]. **Figures 3** and **4**.

Upon close examination, the main axis can be differentiated from the secondary ones with few branches with spines in distal areas. Pneumatophores (1.4–2.9 mm diam.) are more abundant than the oblong blades, without a spine. Always sterile, without receptacles.

Distribution in Eastern Americas [55, 62]: North America: Bermuda, Florida, Louisiana, Massachusetts, Mexico (Gulf of Mexico and Mexican Caribbean), North Carolina, Texas, Virginia. Central America: Belize, El Salvador, Panamá. Caribbean Islands: Bahamas, Barbados, Caribbean, Cuba, Hispaniola, Jamaica, Lesser Antilles, Martinique, St. Eustatius, Virgin Islands. South America: Colombia.

Sargassum natans (Linnaeus) Gaillon, 1828: 355 [63]. **Figures 5–8**.

Homotypic synonym:

Fucus natans Linnaeus, 1753: 1160 [53] (lectotype locality: "Indica", probably Jamaica fide [54, 64].

Heterotypic synonym:

Fucus baccifer Turner, 1802: 55–60 ("*bacciferus*") [65].

Figure 6.
Sargassum natans type I (Puerto Morelos). Scale bar: 1 cm.

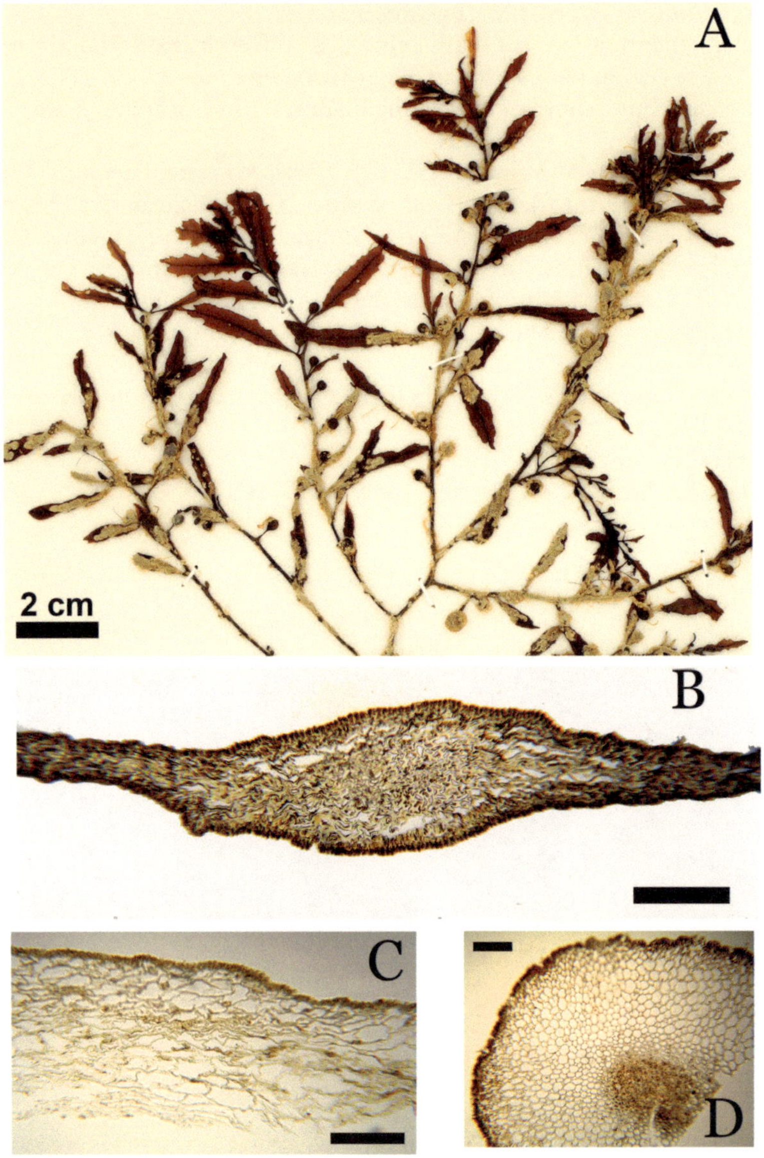

Figure 7.
Sargassum natans type VIII. A: Smooth branches with epibions, spineless pneumatophores and wider blades (NI-MEXU 3604). B: Cross section of the blade the midrib is not observed. C: Cross section of the pneumatophore formed by a very thick layer of cells. D: Cross section of the stem with a very reduced midrib. Scale bars: 200 µm.

Pelagic species branching in several directions, of variable size up to up to 60 cm; usually 20–30 cm wide. Main axis absent, 0.5–2 mm thick, without spines, and with a wiry appearance; in cross-section, in the center, a midrib is observed, surrounded by subspherical cells and a cortex of a layer of quadrangular cells (**Figures 5A, 6, 7A,B**). Blades firm, linear, or lanceolate with the apices pointed, 1–4 cm long by 1–5 mm wide, serrated margin with elongated and acute teeth; in cross-section, a conspicuous midrib (329 µm thick) is observed in the center surrounded by few rounded cells and with a cortex of a layer of quadrangular cells (**Figures 5A,B, 7A,B**); cryptostomatas absent. Blades relatively sparse; generally, a gradual reduction in the secondary branches' size is presented. Pneumatophores abundant, spherical in shape, 2–6 mm in diameter with a 2–5 mm pedicel, with or

Figure 8.
Sargassum natans type VIII (Puerto Morelos). Scale bar: 1 cm.

without spine (**Figures 5A**, **6**, **7A**, **8**) with branching alternate, or at the base and a small leaf-shaped projection; in cross-section, a thick layer of round cells with a hollow center and a cortex of a single layer of cells is observed (**Figures 5C**, **7C**). Fertile specimens never found.

Sargassum natans Type I [8]. **Figures 5** and **6**.

The relative ratio of blades/pneumatophores is 2: 1, pneumatophores 2.1–3.6 mm diameter with frequently a distal spine. Blades linear to narrowly lanceolate, 1–40 mm long, and 1–3 mm wide.

Sargassum natans Type VIII [8]. **Figures 7** and **8**.

Pneumatophores numerous, 2–5 mm in diameter and larger than Type I, with a pedicel (1–4 mm), spine absent and central axis evident. Lanceolate blades are wider than Type I.

Distribution in Eastern Americas [55, 62]: North America: Florida, Louisiana, Maritime Provinces, Massachusetts, Mexico (Gulf of Mexico and Mexican Caribbean), Newfoundland, North Carolina, Texas, Virginia. Central America: Belize. Caribbean Islands: Bahamas, Barbados, Caicos Islands, Caribbean, Cuba, Hispaniola, Jamaica, Lesser Antilles, St. Eustatius. South America: Brazil, Colombia.

The features to identify *Sargassum fluitans* (including Type III) specimens correspond to a branched thallus, with lanceolate or linear blades, with an irregularly toothed margin, slightly oblong pneumatophores and an axis with mainly apical spines; this corresponds to Parr's [8] description. There is an incongruity with Børgesen's original description of material from the Sargasso Sea [7] with large and clearly visible cryptostomatas. Taylor [50] described, however, the absence or presence of only a few cryptostomatas and pneumatophores without a distal spine. Littler and Littler [59] also indicated few or no cryptostomatas. We did not observe cryptostomatas in Mexican material, as was also found in other studies of Caribbean material [66, 67]. Another characteristic that coincides with [50, 59] is the prominent midrib of the blades of the specimen in the MEXU herbarium (**Figures 3A-C**).

In the Mexican material, *S. natans* differ from *S. fluitans* by the absence of spines on the axis. The blades are linear or lanceolate, and the margin is serrated with elongated teeth, cryptostomatas absent, spherical pneumatophores with a pedicel, and often a thorn present depending on type I. Although Taylor [50] and Littler and Littler [59] agreed with almost all the characteristics of *S. natans*, they mentioned that the midrib is not prominent and the pneumatophores have a long spine, coinciding with the material of MEXU corresponding to *S. natans* type I together with the linear blades. *S. natans* type VIII has a thallus also free of spines on the axis and the pneumatophores. It has broader blades with a lanceolate shape, and it is consistent with Taylor [50], without a prominent midrib (**Figures 7B**, **8**). In the MEXU material, the pneumatophores wall is slightly thinner in *S. fluitans* (231–318 μm, **Figure 3D**) than in *S. natans* (362–392 μm, **Figures 5C**, **7C**). In both species, the receptacles are unknown.

2.3 Molecular biology

The identification of *Sargassum* species often requires both extensive morphological studies of specimens and molecular analyses. For molecular identification of *Sargassum* species, the use of genetic DNA markers such as nuclear ITS-2, a portion of the partial RubisCO operon or *rbc*L chloroplast, the variable intergenic mitochondrial spacer mtsp, and the universal mitochondrial markers COI and COX3, have been proposed. However, these barcode markers are less efficient when intraspecific variability and interspecific divergence overlap, as is often the case for *Sargassum* spp. [68]. So far, nuclear ITS-2 and partial plastid RubisCO have been ineffective to identify morphologically different species, but mitochondrial markers (mtsp, COI, and cox3) may have more potential for this purpose [69].

In the Mexican coasts of the Atlantic Ocean, 16 species of *Sargassum* [70] have been reported, and 24 species have been identified for the region from North Carolina to Brazil [62]. However, according to recent molecular studies on species diversity in Mexico, the number of species is lower than initially thought [70]. *Sargassum* species diversity is likely to be overestimated [56]. A recent phylogenetic analyses using three independent molecular markers (COI-5P, ITS-2, and *rbc*L) for Mexican species found that ten previously reported species (*S. bermudense*, *S. buxifolium*, *S. cymosum*, *S. filipendula*, *S. furcatum*, *S. hystrix*, *S. polyceratium*, and *S. vulgare*) were grouped into a single polytomy, with low genetic diversity [29]. Surprisingly, the two pelagic species, *S. fluitans* and *S. natans*, were also included in this polytomy. However, further studies on authentic *Sargassum* materials (types), using higher resolution markers, are needed to validate such a taxonomic proposal. Species diversity has not yet been assessed using a multigene molecular approach, where a higher resolution would be expected using concatenated sequences [71]. Integrative investigations of morphology, life-cycle, and molecular analyses are essential to better understand the taxonomy of *Sargassum*.

2.4 Reproduction

As all species in the genus *Sargassum*, sargasso has a diploid thallus, and clonal reproduction is the only mechanism of propagation known for the sargasso species until today. Neither *S. fluitans* or *S. natans* have been seen to multiply sexually. Vegetative reproduction occurs by fragmentation of old thalli sections (with epibionts) that disintegrate breaking apart, and the newly formed fragments grow again [68]. Generally, no clearly obvious main axes are observed as is the case in benthic species attached to a substrate.

Moreira and Suárez [72] found beach-cast specimens of *Sargassum* spp. in Cuba having fertile receptacles with male or female sexual structures; however, they did not present photographs, and from the illustrations it may be deduced that the thalli were partly decayed, making difficult the identification of these specimens to species level or their habits (pelagic or benthic); however, it is worthwhile to follow up this observation and confirm the identidy of the fertile specimes. Only *S. fluitans* presents occasional cryptostomatas; as Simons [73] indicated, cryptostomatas, although sterile, are homologous to conceptacles (reproductive structures).

In benthic sargasso species, female oogonia are exposed outside the conceptacle through the ostiole, remaining attached to the receptacle, which is a modified terminal structure of the thallus that contains the conceptacles, whose function is to produce reproductive cells. The conceptacle is a cavity (crypt) immersed in the surface of the thallus with an ostiole (opening) to the outside that contains the reproductive structures or gametangia retained by a mucilaginous stalk. Once the sperm is chemically attracted to the oogonium and fertilization occurs, the zygote is released from the receptacle driven by light and temperature cues [74]. The young thalli of benthic *Sargassum* species come into contact with a solid substrate, on which they will produce rhizoids and growing axes. Until now the absence of sexual reproduction in pelagic *Sargassum* species remains an enigma, but the drifting thalli are perfectly capable of growing vegetatively through fragmentation when they are transported by the marine currents and winds [75].

2.5 Distribution, ecology, and origin

The genus *Sargassum* is distributed in all the world's oceans, except for Antarctica [68]. The highest concentration of sargasso (pelagic *Sargassum* species) used to be in the Sargasso Sea, in the subtropical clockwise circulating gyre in the North Atlantic, delimited by the Gulf Stream on the western edge, the North Atlantic Current in the north, the Canary Current in the east and the North Atlantic Equatorial Current in the south. Satellite images have allowed for easier and more comprehensive ways to track the sargasso distribution (e.g. [76, 77]). The accumulations of the sargasso in the Sargasso Sea can be massive (hundreds square meters), completely segregated, distributed in small patches (several to tenths of square meters), or along lines due to the Langmuir circulation [78]. Small or occasionally larger quantities of sargasso from the Sargasso Sea have always been arriving intermittingly to the coasts of the Mexican Caribbean and the Gulf of Mexico when, due to high-pressure anomalies, the algae from this sea were transported southwards, and subsequently introduced into the Caribbean through the Windward, Mona and Anegada Passages, caught up by the Yucatan Current and then re-entering the Sargasso Sea directly through Florida Straits or after passing through a loop in the Gulf of Mexico (The Sargasso Loop System) [79]. Sargasso moves through this large area from spring and early summer, heading towards the Sargasso Sea during autumn and early winter just north of the Bahamas [80]. In certain years, massive quantities of sargasso have beached in the northern Gulf of Mexico, especially in Texas, during the summers [80]. But, in 2011, a new area of concentration was found in the southern tropical Atlantic near the Equator, i.e. Northern Equatorial Recirculation Region (NERR) [81]. Since then, sargasso has been introduced intermittingly into the southern Caribbean, to be transported northwards by the Caribbean and Yucatan currents. This new area of concentration of sargasso in the NERR, from the eastern coast of Africa to Brazil, throughout the Caribbean, and into the Gulf of Mexico has been named the Great Atlantic Sargasso (*Sargassum*) Belt (GASB) by Wang and collaborators [77]. In the peak month of June 2018, the GASB covered 6000 km^2 with an estimated 20 million tons of algal mass, making this the largest macro-algal bloom ever recorded [77]. The algal masses in the GASB show large interannual variability, which to date has been difficult to predict [77]. The influx of sargasso into the Caribbean shows a seasonal pattern, as the North Equatorial Counter Current breaks down from January until May; thus, the generated westward surface flow transports sargasso into the Southern Caribbean Sea [77, 82]. Once released from the NERR and transported into the Caribbean, sargasso flows through the Caribbean and (possibly) the Gulf of Mexico to end its journey in the Sargasso Sea [83]. The pelagic masses of sargasso arrive on the Mexican coasts approximately 2–3 months after their introduction into the southern Caribbean. Southeastern trade winds transport the sargasso masses accumulated in the Yucatan current towards the Mexican coast inundating the beaches with algae at seasonal intervals from March/April until August/September (**Figure 9**) [37, 84].

Sargasso exhibits higher growth rates at higher temperatures (until 30°C) [85] and cannot survive in waters below 18°C [8, 86]. Sargasso in the Sargasso Sea usually has low productivity and a bright yellow color [8], typical of nutrient-depleted populations. Neritic populations (closer to the coast) have greater availability of nutrients (mainly N and P) and develop a deeper brown color, attaining higher photosynthetic capacity (Pmax) and productivity; alkaline phosphatase activity is lower in these nutrient-enriched algae compared to those from the Sargasso Sea, and their tissue

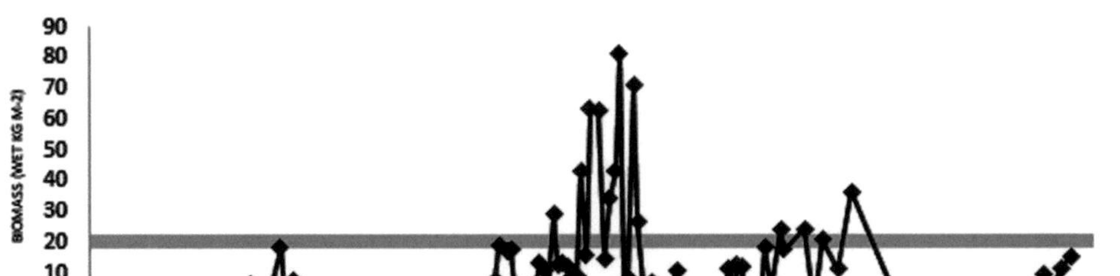

Figure 9.
Biomass of beached fresh sargasso on a beach at Puerto Morelos in the north of the Mexican Caribbean between the summers of 2016 and 2020. The horizontal gray line indicates the minimal biomass when Sargasso-brown-tides (Sbt) are observed (modified from [37]).

concentrations of N and P are higher [87]. Sargasso in the Sargasso Loop System (from the Sargasso Sea to northern Caribbean, and back to th Sargasso Sea directly or through the Gulf of Mexico, see above) acquires nutrients when it passes through the nutrient-rich Gulf of Mexico; thus, there appears to be a neritic-oceanic coupling in this loop system that could have facilitated the adaptation of sargasso to large differences in nutrient availability, maintaining population in oligotrophic waters but rapidly responding to increasing nutrients when available. This capacity to rapidly respond to increasing nutrients may have contributed to the bloom of these algae once introduced into the more eutrophic NERR (than the Sargasso Sea) in 2010/2011.

The different conditions in the Sargasso Sea and NERR may also partially explain differences in specific composition of the sargasso masses in these regions. Studies from the 1990s to 2015 [73], found that *S. natans* I was the most important species in the Sargasso Sea (87%). However, sargasso in the Atlantic Ocean east of the Antilles and the Caribbean in 2014 and 2015 had different specific composition: *S. natans* VIII dominated the western tropical Atlantic (87.3% wet weight), the eastern Caribbean (95.3% wet weight), and the Antilles Current (92.0% wet weight) [88]. *Sargassum fluitans* III generally predominated with more than 60% of the total wet biomass; *S. natans* VIII decreased gradually from 2016 to 2019 but increased again in the beginning of 2020. *S. natans* I almost absent in 2015–2017, appeared in 2018 when it comprised on average 23% of the total sargasso in 2018, increasing in relative abundance from then onwards 2020 [37]. Garcia-Sanchez et al. [37] proposed that differences in abundance of species and their morphological forms could be explained multiple origins of the sargasso transported onshore, or may reflect variable environmental conditions in the seas where they passed through, since sargasso species have different thermal tolerances and growth rates.

In the open ocean, pelagic sargasso represents a diverse community, and it is a critical habitat recruitment area for macrofauna at various development stages, with a complex trophic network of energy flows among herbivores, predators, and detritivores [89]. Sargasso rafts providing breeding and development area for various fish of ecological and commercial interest [90]. For example, the larval fish of the European eel (*Anguila anguila*) and the American eel (*Anguilla rostrata*) hatch and grow in the Sargasso Sea to travel to respectively the European and American continent as juveniles, to return to the Sargasso Sea later in life as sexually mature adults to spawn [91]. Sea turtles are transported by ocean currents and eventually reach sargasso mats providing them with shelter and food [92]. The Sargasso Sea has at least ten species of invertebrates and two vertebrates that are endemic to the Sargasso Sea, including the sargasso fish (*Histrio histrio*), and the sargasso nudibranch (*Scyllaea pelagica*) [90, 93]. Floating masses of sargasso in the Gulf of Mexico are home to of 33 species of macrofauna, including *Callinectes sapidus*, *Latreutes fucorum*, *Portunus sayi*, *Portunus spinicarpus*, *Mugil* sp. and *Balistes capriscus*, and they are considered of importance for the fisheries in the Gulf region [94]. On the other hand, there is evidence that certain deep-sea fish and invertebrates consume the remains of sargasso and associated epibionts exported to the bottom; thereby contributing to the maintenance of deep-sea communities [95]. The superficial influence of the wind causing the Langmuir circulation leads to the formation of algal rafts in the form of rows. Langmuir circulation can sink the algae until a depth (up to 8-10 m), where the sargasso may maintain neutral buoyancy; as the algae age or accumulate epibionts these subsurface masses may sink following the benefits of sargasso, until reaching a depth with higher pressure that causes the implosion of the pneumatophores, resulting in total sinking to the deep-sea floor [95]. The sunken sargasso contains large amounts of carbon, and may thus be an important sink of organic carbon, helping to mitigate global climate change [96].

Beach-cast sargasso, in small quantities, provides food for various coastal species such as amphipods which in turn they are food for birds; it enhances stability to dunes, and it prevents beach erosion [89]. Massive obnoxious quantities of beach cast sargasso have been recorded intermittingly in Texas since the 1890s [97]. Although the recent accumulation of sargasso in the Great Atlantic Sargasso Belt, since 2011, is a new phenomenon, these algae have been previously mentioned (since 1929) [98] in the floristic list from Brasil, the Caribbean islands, and mainland Caribbean [99]. Beach-cast macroalgae, including red, brown, and green algae, with species of the genus *Sargassum* often being the most frequent and abundant, have been frequently observed in the coastal areas of the Mexican Caribbean [100]; thus, beaching of sargasso are not of recent occurrence in this region (**Table 1**). However, the quantities of beach cast sargasso in the Caribbean used to be small or moderate; whereas since 2011 (and since 2015 in Mexico), they can be massive at times.

3. Golden and Sargasso Brown tides

Floating sargasso in the open ocean is considered a valuable habitat (see above), and has also been named "the golden floating rainforest of the Atlantic Ocean" by Lafolley and collaborators [101]. Smetacek and Zingone [96] used the name "Golden Tide" for sargasso; which reflects this value and the yellow brown color of the floating sargasso masses in the oligotrophic open ocean. As indicated above, the "golden forest" in the Sargasso Sea has likely existed for very long time; long enough for endemic species to evolve and species such as the European and American eels to adapt their life cycle to its persistence through time. Although large masses of sargasso from the Sargasso Sea has been arriving at intervals to the US coast of the Gulf of Mexio, in the 1990s, agricultural fertilizers and other pollutants in the USA were linked to unusual accumulations of sargasso biomass at the Mississippi River's mouth between Florida and Texas [87].

The large accumulation of sargasso in the Great Atlantic Sargasso Belt, first reported in 2011, is only a decade old [96]. Sargasso is not new for the southern tropical Atlantic as indicated by the above-mentioned historical reports, but its sudden increase or bloom is a new phenomenon. It is thought that this may have been caused by a combination of various events, all related to human interference with the plant's biogeochemical cycles. It has been attributed to climatological changes related to the sea surface temperature (SST) and an anomaly in North Atlantic Oscillation (NAO) during 2009–2010, which may have introduced large "seed populations" of sargasso into the African side of the NERR [89]. The NERR naturally has more nutrients and is warmer than the Sargasso Sea, providing a more favorable environment for algal growth. Similar to the blooms of other algal species occurring in other parts of the world [102], the current reasoning is that these blooms are associated with an increase in nutrients input into the sea. Upwellig patterns along the African coast have been changing, bringing more nutrients into this oceanic region [77]. Also, increasing Sahara dust storms may have increased nutrients in the NERR [103], as well as increasing nutrient load from river discharges, such as the Congo in Africa or the Amazon in Brazil [77]. Sargasso has higher productivity in neritic waters, rich in nutrients, than, in oceanic waters, deficient in nutrients [88]; in neritic waters, sargasso doubled its biomass in 11 days in contrast to the 50 days in oceanic waters [104]. In addition, sargasso has no physical barriers for expansion, no competitors, and no large herbivores.

Mexico started to receive unusual quantities of sargasso late 2014, and the first massive beachings were reported in 2015 [4]. A general overview of the massive influx of sargasso into Mexico has been presented recently by Chavez and collaborators [84]. The first significant influx of sargasso was in the summer of 2015, when on average 319 m^3 of sargasso were removed per km of beach per day in the northern part of the Mexican Caribbean [4]. Satellite images of sargasso in front of the Mexican coast revealed large interannual fluctuations in abundance, reaching a maximum cover of 22,900 ha in September 2018 [84]. Generally, large abundance in the ocean corresponded with larger quantities on the beaches, although not all coasts were equally impacted. During the peak years of 2018 and 2019, on average 3.2×10^3 and 1.7×10^3 m^3 were collected per month, per km of beach in the Northern Mexican Caribbean, respectively [84].

The excessive amount of sargasso on the beaches generates problems for productive sectors such as tourism. Decaying sargasso covers the beaches with a brown mud; its decomposition emits an unpleasant smell that attracts insects and can cause serious health problems. The business sector and the local, state, and federal governments are concerned about the consequences of the damage the emblematic image of these pristine turquoise beaches would bring to the tourist activity in the region [2]. Economic parameters such as the billing rate, jobs, and the hotel activity in this region, did not detect a discernible impact on the state's economies during 2018 and 2019 [2]. However, hotel occupancy has declined and some hotels have invested major efforts in maintaining their beach fronts free from sargasso [84].

Massive beaching of sargasso are creating havoc to Caribbean coastal ecosystems as leachates and particulate organic matter from the stranded decaying algal masses deplete oxygen, reduce light and deteriorate water quality, which leads to the mortality of nearshore seagrasses and fauna, interference with seaward journeys of juvenile turtles, enhance beach erosion, and changes in trophic dynamics of benthic organisms. Excessive nutrient concentrations caused by sargasso leachates have been measured until the reefs, and release of heavy metals sequestered by sargasso (such as arsenic) are also of concern (summarized in [84]). This is why van Tussenbroek and collaborators [99] suggested the term Sargasso (*Sargassum*) Brown Tide instead of Golden Tide; to refer to the decaying algal masses washed ashore, coloring the near-shore waters with a dark, murky brown color.

4. Uses of sargasso and perspectives

The oldest known record of the use of sargasso was that of the Portuguese physician and naturalist Cristóbal de Acosta (1515–1594). Acosta, in his "Treatise on drugs and medicine of the East Indies" of 1578, described the sargasso located between 18 and 30° N, that is, in the eastern part of America. He noted that this hearty herb was pickled and flavored with fennel. He claims that he fed it raw and cooked to a sailor with "bad urine" [urinary tract infections] and claimed that he did it well and the sailor took it to his house when they disembarked [19]. In 1750, Georg Eberhard Rumphius (1627–1702), a Dutch naturalist famous for his "Herbarium amboinense", described sargasso with narrow and long leaves, without roots, which are boiled and drunk with great success against the water accumulated in the kidneys. [105]. In his "Dictionnaire raisonné universal d'histoire naturelle", Jacques Christophe Valmont de Bomare (1731–1807) described sargasso from the west coast of Africa in 1764 and identified it as *Fucus natans*, which is eaten in salads, he also uses it to facilitate delivery, it is used against urine retention and against scurvy [106]. In 1771, Dr. Vicente de Lardizábal Dubois (1746–1814) in his treatise on sargasso "The consolation of the navigators", called it sea lentil or watercress that served as food for birds and cattle on boats and as beneficial medicine against scurvy [107]. Years later, Dr. Antonio Corbella published a dissertation where he did not recommend sargasso for scurvy, which could be better cured with a traditional vegetarian diet and lemon juice [14]. Sargasso was mentioned in 1799 by José de

Viera y Clavijo (1731–1813), who presented morphological descriptions and characteristics of its habitat; he cited the uses published by Lardizábal. The works of Viera and Clavijo were published in the 19th century [22, 23].

In the 20th century and the beginning of the 21st, many research efforts and technological developments of sargasso have been developed. Sargasso can provide sustainable alternative routes based on renewable raw materials that can provide biofuels (biofuels, biomethane, nanocarbon), obtaining chemicals such as sugars, proteins, and alginates and fucoidans. Also purchase pharmaceutical products (antiproliferative, antiprotozoal, antioxidant, hepaprotective, antileishmanial), fertilizers, cosmetics (for hair treatments), bioplastics, biopolymers, cellulose, and other products such as materials to make shoes, bricks, varnishes, paper, cardboard, as well as services on environmental impacts (leachates) and on bioremediation.

Table 2 summarizes the uses that have been implemented in contemporary Mexico. Sargasso is not feasible for the production of biodiesel due to low concentrations of lipids and fatty acids, so other biofuels have been explored through the fermentation and anaerobic digestion of sargasso biomass to obtain liquid fuels such as ethanol or biogas such as biomethane [109]. The Renewable Energy Unit, Centro de Investigaciones Científicas de Yucatán AC (CICY), investigates sargasso with a fungal pretreatment that produced a 20% increase in methane yield [112].

Alginates are phycocolloids present in all brown algae, including sargasso. *S. fluitans* from the Caribbean, particularly in Cuba and Puerto Rico, with alginate yields of 16–19% [122, 123]. Sargasso yields are lower than those of other brown algae such as *Macrocystis pyrifera*, yielding 26.5 to 35.9% [124]. In Mexico, the Alquimar industry has two patents on alginate extraction technologies from sargasso (**Table 2**).

Some bioactive molecules to combat diseases caused by protozoa or substances that have a hepaprotective or antioxidant nature are under investigation, but with good results for the near future [119–121].

The Salgax company, among other companies, already has sargasso based fertilizers for sale with a sale price of $ 24 US for a 5-liter jug, in addition to other cosmetic products [111]. Other companies manufacture footwear and bricks with a majority percentage of sargasso [109, 115], among other products (**Table 2**).

Some research centers focus on bioremediation and leachate control processes such as the Center for Applied Physics and Advanced Technology (CFATA) and Autonomous National University of Mexico (UNAM) [109] or, in the case of biopolymer development, the Open and Distance University of Mexico of Quintana Roo [114].

However, despite the effort that has been achieved in the region of Mexico there are still some characteristics of sargasso that can hinder or interfere with some uses. Algae are known to be excellent at absorbing heavy metals and other pollutants, particularly arsenic [38]. Sargasso from the Mexican Caribbean has high arsenic concentrations (24–172 ppm), depending on the region. *Sargassum natans* VIII showed the highest concentration (123 ppm) and in the lowest concentration *S. fluitans* III (59 ppm) and *S. natans* I (55 ppm); however, in all morphotypes, arsenic exceeded the allowed limits (40 ppm DW). Arsenic concentrations are a concern both for environmental and aquifer contamination and for its use as food or biofertilizer [38]. Further studies are required to find out which arsenic speciation is more toxic (inorganic or organic arsenic such as arsenate and arsenite) and to investigate which parts of sargasso are more likely to absorb arsenic (stems, pneumatophores) [125]. Apart from arsenic, sargasso is high in salt and ash. Salt removal is expensive as it requires the use of large amounts of fresh water.

Sargasso is free, but it is expensive to collect and often requires specialized machinery depending on whether it is harvested from the ocean or from the beach. It is also necessary to take into account its storage due to the toxic gases it gives off, such as hydrogen sulfide and ammonia, together with the leachate generated,

Use	End product / Research	Name of company / Research institution	Ref.
Agricultural	Fertilizers and fucoidan extract	Alquimar	[108]
	Substrate for growing mushrooms	Yucatan Scientific Research Center (CICY), Postgraduate College, Puebla Campus & Popular Autonomous University of the State of Puebla (UPAEP)	[109]
	Organic fertilizer and cellulose	Grupo MD	[110]
	Fertilizers	Salgax Biotecnología Marina Aplicada S. de R.L de C.V	[111]
Animal feed	Non-traditional ingredients	National Institute of Medical Sciences and Nutrition Salvador Zubirán (INCMNSZ)	[109]
Biofuels	Biopellets & bioplastic	Energryn / Solesyto	[109]
	Cellulose	Grupo MD	[110]
	Biomethane	Renewable Energy Unit, Scientific Research Center of Yucatán AC (CICY)	[112]
	Biomass (*Sargassum* + organic waste), nanocarbons and extracts	National Technology of Mexico / IT of Cancun	[109]
Bioplastics	Bioplastics	Abaplas	[109, 113]
	Biopolymer	Open and Distance University of Mexico of Quintana Roo	[114]
Bioremediation & purification	Removing contaminants such as metals, sulphates and pigments (phenolic compounds) from water	Center for Applied Physics and Advanced Technology (CFATA), Autonomous National University of Mexico (UNAM)	[109]
	Activated carbon	Transporte Marítima Mexicana (TMM)	[109]
Footwear	Shoes	Renovare	[115]
Construction material	Bricks "Sargablock"	Blue Green (Casa Angelita)	[109]
Cosmetics	Capillary treatment	Salgax Biotecnología Marina Aplicada S. de R.L de C.V	[111]
Environmental impact	Eco-toxological impact of sargasso leachates	Institute of Marine Sciences and Limnology, UNAM	[109]

Use	End product / Research	Name of company / Research institution	Ref.
Paints	Wood varnish	Salgax Biotecnología Marina Aplicada S. de R.L de C.V	[111]
Paper and cardboard products	Paper products & fertilizer	Dianco México	[109, 116]
	Paper, notebooks	Salgax Biotecnología Marina Aplicada S. de R.L de C.V	[111]
		Sarganico Productos sustentables	[117]
Pharmaceuticals & biomedical	Fucoidans	Alquimar	[108]
	Cytotoxic, antiproliferative, and antiprotozoal	Department of Marine Resources, Unidad Mérida, Yucatan (CINVESTAV)	[118]
	Fucoidans: Protective effect against oxidative stress	Hospital de Especialidades, I. García Téllez, National Medical Center, Mexican Institute of Social Security (IMSS)	[119]
	Hepatoprotective effect of a fucoidan extract	Hospital de Especialidades, I. García Téllez, National Medical Center (IMSS)	[120]
	Antileishmanial	CINVESTAV	[121]
Phycocolloids	Alginate, fucoidans & fertilizer	Alquimar	[108]

Table 2.
*Current uses of pelagic sargasso of Mexico (**Sargassum** spp.)*

which can contaminate groundwater. As we can see, there are still enormous challenges to overcome. However, developing an ecological pretreatment method by incorporating the use of low-cost ionic resins to remove contaminants and promote cell fracture by solubilizing the polymers contained in sargasso would be methods that could be explored in the future and also the biorefineries concept; based on sargasso in Mexico, which would produce a feasible techno-economic model for the region (Paula Sánchez. Com. per.).

Creating a political or governance framework to manage the influx of sargasso and standards for the comprehensive management of sargasso and the availability of funds will be challenged to achieve in the future. It is our duty to encourage entrepreneurship, incorporating the newest scientific and technological advancements..

5. Conclusions

- Historical records of sargasso go back to centuries XV to XXI, not only in the Sargasso Sea but also throughout the Caribbean. Although there is also a pre-Hispanic record

- The taxonomy of the genus *Sargassum* and the position of the pelagic *Sargassum* species (sargasso) still require further research.

- The absence of sexual reproduction is an enigma; further life-cycle studies and population genetic studies may shed further light on it.

- In the past sargasso has been used to treat urinary tract problems, to facilitate delivery, and also as food; it may be worthwhile to follow this up with further research.

- Sargasso in the open ocean is a Golden Tide, sustaining and providing energy for a diverse community, and serving as a nursery or refuge for commercially important or iconic species.

- Sargasso may be an important sink of organic carbon, helping to mitigate global climate change.

- The sargasso blooms in the NERR (North Equatorial Recirculation Region) since 2011, creating the Great Atlantic Sargasso Belt, are likely the new normal condition for the region.

- Likely causes of this bloom are thought to be (a combination of) changes in prevailing winds and currents due to Climate Change and increasing nutrient into the NERR from changes in upwelling and land-runoff.

- Sargasso, although capable to persist in the oligotrophic Sargasso Sea, also has the capacity to rapidly respond to increasing nutrients that may have contributed to the bloom of these algae in the NERR

- Massive quantities of beached sargasso create the Sargasso Brown Tides causing major havoc to ecosystems and local economies.

- Better understanding of the species, the movements of the rafts in the ocean, and new research into valorization, are necessary to create novel strategies of adaptation to the recent blooms of sargasso.

- Mexico has been working on the following mitigation actions: strengthening of monitoring programs, removal of oceanic and coastal sargasso, regulation plan for the deposit of sargasso, creation of a monetary fund, and increasing integration of national knowledge and capacities.

Acknowledgements

The authors thank Dr. Michael J. Wynne for reviewing the manuscript, Martha García Sánchez for providing us with the live sargasso images. Thanks to Dr. Nina Lundholm for providing Liebmann's herbarium specimens from the Natural History Museum of Denmark.

Author details

José Luis Godínez-Ortega[1*], Juan V. Cuatlán-Cortés[1], Juan M. López-Bautista[2] and Brigitta I. van Tussenbroek[3]

1 Institute of Biology, National Autonomous University of Mexico, Mexico City, Mexico

2 Department of Biological Sciences, The University of Alabama, Tuscaloosa, AL, USA

3 Institute of Marine Sciences and Limnology, Universidad Nacional Autónoma de México, Puerto Morelos, Quintana Roo, México

*Address all correspondence to: jlgo@unam.mx

References

[1] Fidai YA, Dash J, Tompkins E, Tonon T. A systematic review of floating and beach landing records of *Sargassum* beyond the Sargasso Sea. Environmental Research Communications. 2020;2:122001 DOI: 10.1088/2515-7620/abd109

[2] Espinoza LA, Li JJ. El riesgo del sargazo para la economía y turismo de Quintana Roo y México. BBVA Research. 2020;20:2-33.

[3] Robledo D, Vázquez-Delfín E. Sargazo, conociendo al "enemigo". Revista Avance y Perspectiva. 2019; 5(3):1-7.

[4] Rodríguez-Martínez R, Tussenbroek B, Jordán-Dahlgren E. Afluencia masiva de sargazo pelágico a la costa del Caribe mexicano (2014-2015). In: García-Mendoza E, Quijano-Scheggia SI, Olivos-Ortiz A, Núñez-Vázquez EJ, editors. Florecimientos algales nocivos en México. 1st ed. Quintana Roo: CICESE; 2016. p. 352-365. ISBN: 978-607-95688-5-6

[5] Jerzmańska A, Kotlarczyk J. The beginnings of the Sargasso assemblage in the Tethys? Palaeogeography, Palaeoclimatology, Palaeoecology. 1976;20(4):297-306. DOI:10.1016/0031-0182(76)90009-2

[6] Yip ZT, Quek RZB, Huang D. Historical biogeography of the widespread macroalga *Sargassum* (Fucales, Phaeophyceae). Journal of Phycology. 2020;56:300-309. DOI: 10.1111/jpy.12945

[7] Børgesen F. The marine algae of the Danish West Indies. Part 2. Phaeophyceae. Dansk Botanisk Arkiv. 1914;2(2):1-68.

[8] Parr AE. Quantitative observations on the pelagic *Sargassum* vegetation of the western North Atlantic. With preliminary discussion of morphology and relationships. Bulletin of the Bingham Oceanographic Collection. 1939;6:1-94.

[9] Pérez-Rubin JF. Las algas y los antiguos navegantes españoles (1492-1792). Revista del Instituto Español de Oceanografía. 2016;24:62-75.

[10] Diccionario Maya Cordemex. Maya-Español, Español-Maya. Mérida: Ediciones Cordemex. 1980; 984 p.

[11] Colón C. Relaciones y Cartas. Madrid: Librería de la Viuda de Hernando y Ca.; 1892. 423 p.

[12] Hydrography: Maritime paper, Reviews of voyages. I. Sargasso weed. In: The Nautical Magazine for 1832. The Nautical Magazine. 1832;June: 175-179

[13] Solano-Constâncio F. Novo Diccionario critico e etymologico da Lingua Portugueza. Paris: Na Officina Typographica de Casimir; 1836. 976 p.

[14] Cabral JP. A apanha de algas na ilha da Ínsua (Caminha) nos séculos XVII - XIX. Singularidades e conflitos. Finisterra. 2005;40(80):5-22. DOI:10.18055/Finis1475

[15] Oliveira EV, Galhano F. Apanha do sargaço no norte de Portugal. Trabalhos de Antropologia e Etnologia. 1958;16(1-4):63-170.

[16] Krüger F. Apuntes etnográfico-lingüísticos de Póvoa de Varzim. Boletín de Filología. 1936;4(1-2):119-120.

[17] Oliveira VE, Galhano F, Pereira B. Actividades agro-marítimas em Portugal. Lisboa: Etnográfica Press; 1990. 236 p.

[18] Fernández De Oviedo y Valdés G. Historia general y natural de las Indias,

islas y tierra-firme del mar océano. Madrid: Imprenta de la Real Academia de la Historia; 1851. 774 p.

[19] Acosta C. Tractado de las drogas, y medicinas de las Indias Orientales, con sus Plantas debuxadas albiuo. Burgos: Martín de Victoria impressor de su Magestad; 1578. 448 p.

[20] Humboldt A. Cristóbal Colón y el descubrimiento de América: historia de la geografía del Nuevo Continente y de los progresos de la astronomía náutica en los siglos XV y XVI. Tomo II. Madrid: Librería de la Viuda de Hernando y Cª; 1892. 382 p.

[21] Ruiz López H. De vera fuci natantis fructificatione. Commentarius. Matriti: apud viduam e filium Petri Marin; 1798. 38 p.

[22] Viera y Clavijo J de, Diccionario de Historia natural de las Islas Canarias, ... Tomo I. Las Palmas; 1866. 344 p.

[23] Martin Aguado M, 1957. Las algas de Canarias en la obra científica de Viera y Clavijo. Revista de historia canaria. Facultad de Filosofía y Letras de la Universidad de La Laguna. 1957;23(117-118):6-52. https://mdc.ulpgc.es/cdm/ref/collection/revhiscan/id/178

[24] Monge F, Olmo M del. Las Noticias de Nutka de José Mariano Moziño. Madrid: Ediciones Doce Calles, Consejo Superior de Investigaciones Científicas, 1990; 266 pp.

[25] Liebmann FM. Beretning om Amerikas Vandplanter af Algernes Classe. Overs. Kongel. Danske Vidensk. Selsk. Forh. Medlemmers Arbeider 1846;72-77.

[26] Taylor WR. Tropical marine algae of the Arthur Schott Herbarium. Publ. Field Mus. Nat. Hist., Bot. Ser. 1941;20:87-104.

[27] Martínez Lozano S, Guajardo Ríos O. Lista sistemática de las algas marinas del Puerto El Mezquital, Matamoros, Tamaulipas, México. Biotam (México). 1991;3(3):16-26.

[28] Martínez Lozano S, López Bautista JM. Algas marinas bénticas de Soto la Marina, Tamaulipas, México. Publ. Biol. Fac. Ci. Biol., Univ. Autón. Nuevo León (México). 1991;5(2):13-22.

[29] González-Nieto D, Cabral-Oliveira M, Núñez-Resendiz M, Dreckmann K, Mateo-Cid LE, Sentíes A. Molecular assessment of the genus *Sargassum* (Fucales, Phaeophyceae) from the Mexican coasts of the Gulf of Mexico and Caribbean, with the description of *S. xochitlae* sp. nov. Phytotaxa. 2020;461(4):254-274. DOI: 10.11646/phytotaxa.461.4.3

[30] De la Campa de Guzmán S. Notas preliminares sobre un reconocimiento de la flora marina del Estado de Veracruz. Anales Inst. Nac. Invest. Biol.-Pesq. 1965;1:9-49.

[31] Huerta L, Chávez ML, Sánchez Rodríguez ME. Algas marinas de la Isla de Enmedio, Veracruz. En: Mem. Congr. Nac. Oceanogr. (Guaymas, Sonora, México, 1974). 1977;5:314-325.

[32] Lehman RL, Tunnell JW, Jr. Species composition and ecology of the macroalgae of Enmedio Reef, Veracruz, Mexico. Texas J. Sci. 1992;44(4):445-457.

[33] Huerta L, Mendoza González AC, Mateo Cid LE. Avance sobre un estudio de las algas marinas de la Península de Yucatán. Phytologia. 1987;62(1):23-53.

[34] Huerta L. Contribución al conocimiento de las algas de los bajos de la Sonda de Campeche, Cozumel e Isla Mujeres. Anales Esc. Nac. Ci. Biol. 1958;9:115 -123.

[35] Kornicker LS, Bonet F, Cann R, Hoskin CM. Alacran Reef, Campeche Bank, México. Publ. Inst. Mar. Sci. 1959;6:1-22.

[36] Kim CS. Marine Algae of Alacran Reef, Southern Gulf of Mexico. [Ph. D. Thesis], Durham: Duke University; 1964.

[37] García-Sánchez M, Grahama C, Vera E, Escalante-Mancera E, Álvarez-Filip L, van Tussenbroek BI. Temporal changes in the composition and biomass of beached pelagic *Sargassum* species in the Mexican Caribbean. Aquatic Botany. 2020;167:103275. DOI: 10.1016/j.aquabot.2020.103275

[38] Rodríguez-Martínez RE, Roy PD, Torrescano-Valle N, Cabanillas-Terán N, Carrillo-Domínguez S, Collado-Vides L, García-Sánchez M, van Tussenbroek BI. Element concentrations in pelagic *Sargassum* along the Mexican Caribbean coast in 2018-2019. 2020;PeerJ 8:e8667. DOI: 10.7717/peerj.8667

[39] Huerta L, Garza Barrientos MA. Contribución al conocimiento de la flora marina de la zona sur del litoral de Quintana Roo, México. Anales Esc. Nac. Ci. Biol. 1980;23:25-44.

[40] Aguilar Rosas MA. Algas marinas bentónicas de la Reserva de la Biósfera de Sian Ka'an, Quintana Roo, México. In: Navarro D, Robinson YJG, editors. Diversidad biológica en la Reserva de la Biósfera de Sian Ka'an, Quintana Roo, México. México: Centro de Investigaciones de Quintana Roo (CIQRO); 1990. p.13-34.

[41] Collado L, Ortegón I, Sentíes A, Comba L, González J. Macroalgae of Puerto Morelos Reef System, Mexican Caribbean Hidrobiológica. 1998;8(2):133-143.

[42] Dreckmann KM, Stout I, Sentíes Granados A. Lista actualizada de las algas marinas bentónicas de Puerto Morelos, Quintana Roo, Caribe Mexicano. Polibotánica. 1996;3:1-17.

[43] Aguilar Rosas M A, Aguilar Rosas L E, Aguilar Rosas R. Algas marinas de la región central de Quintana Roo, México. Polibotánica. 1998;7:15-32.

[44] Mendoza-Becerril MA, Serviere-Zaragoza E, Mazariegos-Villarreal A, Rivera-Perez C, Calder DR, Vázquez-Delfín EF, Freile-Pelegrín Y, Agüero J, Robledo D. Epibiont hydroids on beachcast *Sargassum* in the Mexican Caribbean. PeerJ. 2020; 8:e9795. DOI: 10.7717/peerj.9795

[45] Humm HJ, Hildebrand HH. Marine Algae from the Gulf Coast of Texas and Mexico. Publ. Inst. Mar. Sci.1962;8:227-268.

[46] Earle SA. Benthic algae and seagrass species in the Gulf of Mexico. In: Bushnell VC, editors. Serial Atlas of the Marine Environment. New York: American Geographical Society, p. 25-29.

[47] Murray G. Catalogue of the marine algae of the West Indian region. J. Bot. 1888: 26:193-196, 237-243, 303-307, 331-338, 358-363.

[48] Murray G. Catalogue of the marine algae of the West Indian region. London: Dulau Co.; 1889. 46 p.

[49] Huerta L. Lista preliminar de las algas marinas del litoral del Estado de Veracruz. Bol. Soc. Bot. México. 1960;25:39-45.

[50] Taylor WR. Marine algae of the eastern tropical and subtropical coast of the Americas. Ann Arbor: University of Michigan; 1960. 870 p.

[51] Tsuda RT, Abbott IA. Collection, handling, preservation, and logistics. In: Littler MM, Littler DS, editors. Handbook of phycological methods. Ecological field methods: macroalgae. Vol. 4. Cambridge: Cambridge University Press; 1985. p. 67-86. ISBN 0521249155

[52] Agardh CA. Species algarum rite cognitae, cum synonymis, differentiis specificis et descriptionibus succinctis. Vol. 1. Lund: Berling; 1820. 168 p.

[53] Linnaeus C. Species plantarum, exhibentes plantas rite cognitas, ad genera relatas, cum differentiis specificis, nominibus trivialibus, synonymis selectis, locis natalibus, secundum systema sexuale digestas. Vol. 2. Holmiae [Stockholm]: Impensis Laurentii Salvii; 1753. 1200 p.

[54] Silva PC, Basson PW, Moe RL. Catalogue of the benthic marine algae of the Indian Ocean. California: University of California Publications in Botany; 1996. 1259 p.

[55] Guiry MD, Guiry GM. AlgaeBase, World-wide electronic publication. [Internet]. 2019. Available from: https://www.algaebase.org [Accessed: 2020-12-20].

[56] Mattio L, Payri CE. 190 Years of *Sargassum* Taxonomy, Facing the Advent of DNA Phylogenies. The Botanical Review. 2011;77:31-70. DOI: 10.1007/s12229-010-9060-x

[57] Mattio L, Payri CE, Verlaque M, Reviers B. Taxonomic revision of *Sargassum* sect. *Acanthocarpicae* (Fucales, Phaeophyceae). Taxon. 2010;59 (3):896-904. DOI: 10.2307/25677676

[58] Dixon RRM, Mattio L, Huisman JM, Payri CE, Bolton JJ, Gurgel CFD. North meets south - Taxonomic and biogeographic implications of a phylogenetic assessment of *Sargassum* subgenera *Arthrophycus* and *Bactrophycus*. Phycologia. 2014;53(1):15-22.

[59] Littler DS, Littler MM. Caribbean reef plants. Washington: OffShore Graphics; 2000. 542 p.

[60] McNeill J, Barrie FR, Burdet HM, Demoulin V, Hawksworth DL, Marhold K, Nicolson H, Prado J, Silva PC, Skog JE, Wiersema JH, Turland NJ, editors. International code of botanical nomenclature (Vienna Code): Adopted by the Seventeenth International Botanical Congress Vienna, Austria, July 2005. Ruggell, Liechtenstein: Gantner Verlag. 2006. 568 p.

[61] Børgesen F. The species of *Sargassum* found along the coasts of the Danish West Indies with remarks upon the floating forms of the Sargasso Sea. In: Jungersen HFE, Warming E, editors. Mindeskrift i Anledning af Hundredaaret for Japetus Steenstrups Fødsel, Vol. Art. XXXII. Kobenhavn: Lunos bogtrykkeri; 1914. p. 1-20.

[62] Wynne MJ. A checklist of benthic marine algae of the tropical and subtropical western Atlantic: fourth revision. Nova Hedwigia, Beihefte, Beih. 2017;145:1-202. ISBN 978-3-443-51067-1

[63] Gaillon B. Résumé méthodique des classifications des Thalassiophytes. Dictionnaire des Sciences Naturelles [Levrault]. 1828;53:350-406.

[64] Spencer MA, Linda MI, Jarvis CH. Typification of Linnaean names relevant to algal nomenclature. Taxon. 2009;58 (1):237-260. DOI: 10.1002/tax.581023

[65] Turner D. A synopsis of the British Fuci. London: F. Bush, Yarmouth. 1802. 189 p.

[66] Camacho O, Mattio L, Draisma S, Fredericq S, Diaz-Pulido G. Morphological and molecular assessment of *Sargassum* (Fucales, Phaeophyceae) from Caribbean Colombia, including the proposal of *Sargassum giganteum* sp. nov., *Sargassum schnetteri* comb. nov. and *Sargassum* section *Cladophyllum* sect. nov., Systematics and Biodiversity. 2014; 13(2):105-130. DOI: 10.1080/14772000.2014.972478

[67] Rosado-Espinosa LA, Freile-Pelegrín Y, Hernández-Nuñez E, Robledo D. A comparative study of *Sargassum* species from the Yucatan peninsula coast: morphological and chemical characterisation. Phycologia. 2020;59(3):261-271. DOI: 10.1080/00318884.2020.1738194

[68] Kilar JA, Hanisak MD, Yoshida T. On the expression of phenotypic variability: why is *Sargassum* so taxonomically difficult? In: Abbott IA, editor. Taxonomy of Economic Seaweeds. Vol. 3. La Jolla: Publication of the California Sea Grant College. 1992. p. 95-117.

[69] Mattio L, Payri C. Assessment of five markers as potential barcodes for identifying *Sargassum* subgenus *Sargassum* species (Phaeophyceae, Fucales). Cryptogamie, Algologie. 2010;31(4):467-485. https://sciencepress.mnhn.fr/en/periodiques/algologie

[70] Ortega MM, Godínez JJ, Garduño G. Catálogo de algas bénticas de las costas mexicanas del Golfo de México y Mar Caribe. México: Instituto de Biología, UNAM; 2001. 594 p. ISBN 968369566-3

[71] Huang C, Sun Z, Gao D, Yao J, Hu ZM, Xu K, Chen W. Molecular analysis of *Sargassum* from the northern China seas. Phytotaxa. 2017;319(1):71-83. DOI: 10.11646/phytotaxa.319.1.3

[72] Moreira L, Suárez AM. Estudio del género *Sargassum* C. Agardh, 1820 (Phaeophyta, Fucales, Sargassaceae) en aguas cubanas. 4. Reproducción sexual en *Sargassum natans* (Linnaeus) Meyer y *S. fluitans* Børgesen. Revista de Investigaciones Marinas. 2002;23(1):63-65.

[73] Simons EB. A Morphological Study of *Sargassum filipendula*. Botanical Gazette. 1906;41(3):161-182.

[74] Liang ZR, Wang FJ, Sun XT, Wang WJ, Liu FL. Reproductive Biology of *Sargassum thunbergii* (Fucales, Phaeophyceae). American Journal of Plant Sciences. 2014;5:2574-2581. DOI: 10.4236/ajps.2014.517271

[75] Johns EM, Lumpkina R, Putmanb NF, Smitha RH, Muller-Karger FE, Rueda-Roa DT, Huc C, Wangc M, Brooksd MT, Gramera LJ,Werner FE. The establishment of a pelagic *Sargassum* population in the tropical Atlantic: Biological consequences of a basin-scale long distance dispersal event. Progress in Oceanography. 2020; 182:102269. DOI: 10.1016/j.pocean.2020.102269

[76] Wang M, Hu C. Mapping and quantifying *Sargassum* distribution and coverage in the Central West Atlantic using MODIS observations. Remote Sensing of Environment. 2016;183:350-367. DOI: 10.1016/j.rse.2016.04.019

[77] Wang M, Hu C, Barnes BB, Mitchum G, Lapointe B, Montoya JP. The great Atlantic *Sargassum* belt. Science. 2019;365:83-87. DOI: 10.1126/science.aaw7912

[78] Martin LM. Pelagic *Sargassum* and its associated mobile fauna in the Caribbean, Gulf of Mexico, and Sargasso Sea. [thesis] Texas: Professional Studies of Texas A&M University; 2016.

[79] Frazier J. Advanced Prediction of the Intra-Americas *Sargassum* Season through Analysis of the *Sargassum* Loop System Using Remote Sensing Technology [Master's thesis], Texas: A & M University; 2014. Available electronically from: https://oaktrust.library.tamu.edu/bitstream/handle/1969.1/153840/FRAZIER-THESIS-2014.pdf?sequence=1&isAllowed=y

[80] Gower JFR, King SA. Distribution of floating *Sargassum* in the Gulf of Mexico and the Atlantic Ocean mapped using MERIS. International Journal of

Remote Sensing. 2011;32(7):1917-1929. DOI: 10.1080/2150704X.2013.796433

[81] Gower J, Young E, King S. Satellite images suggest a new *Sargassum* source region in 2011. Remote Sens. Lett. 2013;4:764-773. DOI:10.1080/21507 04X.2013.796433

[82] Franks JS, Johnson DR, Ko DS. Pelagic *Sargassum* in the tropical North Atlantic. Gulf Caribb. Res. 2016;27:SC6-11. DOI:10.18785/gcr.2701.08

[83] Brooks MT, Coles VJ, Hood RR, Gower JFR. Factors controlling the seasonal distribution of pelagic *Sargassum*. Mar. Ecol. Prog. Ser. 2018:599:1-18. DOI: org/10.3354/ meps12646.

[84] Chávez V, Uribe-Martínez A, Cuevas E, Rodríguez-Martínez RE, van Tussenbroek B I, et al. Massive Influx of Pelagic *Sargassum* spp. on the Coasts of the Mexican Caribbean 2014-2020: Challenges and Opportunities. Water. 2020;12(10):2908. DOI: 10.3390/ w12102908

[85] Tussenbroek BI. Environmental data of Puerto Morelos (in process).

[86] Winge O. The Sargasso Sea, its boundaries and vegetation. In: Report on the Danish Oceanographical Expeditions 1908-10 to the Mediterranean and Adjacent Seas. Miscellaneous Papers. 1923;3(2):3-34.

[87] Lapointe B. A comparison of nutrient-limited productivity in *Sargassum natans* from neritic vs. oceanic waters of the western North Atlantic Ocean. Limnology and Oceanography. 1995;40:625-633. DOI: 10.4319/lo.1995.40.3.0625

[88] Schell J, Goodwin D, Siuda A. Recent *Sargassum* Inundation Events in the Caribbean. Oceanography. 2015;28(3):8-11. DOI: 10.5670/ oceanog.2015.70

[89] Uribe-Martínez A, Guzmán-Ramírez A, Arreguín-Sanchéz F, Cuevas E. El sargazo en el Caribe mexicano, revisión de una historia impensable. In: Rivera-Arriaga E, Azuz-Adeath I, Cervantes Rosas OD, Espinoza-Tenorio A, Silva-Casarín R, Ortega-Rubio A, Botello AV, Vega-Serratos BE, editors. Gobernanza y Manejo de las Costas y Mares ante la Incertidumbre. Una Guía para Tomadores de Decisiones. Campeche: Universidad Autónoma de Campeche, Ricomar; 2020. p. 743-768. DOI: 10.26359/epomex.0120

[90] South Atlantic Fishery Management Council (SAFMC). Fishery management plan for pelagic *Sargassum* habitat of the South Atlantic Region. Carolina: South Atlantic Fishery Management Council. 2002. DOI:10.1007/BF00446711

[91] van Ginneken, Vincent JT, Gregory E M. The European eel (*Anguilla anguilla*, Linnaeus), its lifecycle, evolution and reproduction: a literature review. Reviews in Fish Biology and Fisheries 2005;15(4):367-398.

[92] Witherington B, Hirama S, Hardy R. Young sea turtles of the pelagic *Sargassum* dominated drift community: habitat use, population density, and threats. Marine Ecology Progress Series. 2012;463:1-22. DOI: 10.3354/ meps09970

[93] Laffoley Dd'A, Roe HSJ, Angel MV, Ardron J, Bates NR, Boyd LL, Brooke S, Buck KN, Carlson CA, Causey B, Conte MH, Christiansen S, Cleary J, Donnelly J, Earle SA, Edwards R, Gjerde KM, Giovannoni SJ, Gulick S, Gollock M, Hallet J, Halpin P, Hanel R, Hemphill A, Johnson RJ, Knap AH, Lomas MW, McKenna SA, Miller MJ, Miller PI, Ming FW, Moffitt R, Nelson NB, Parson L, Peters AJ, Pitt J, Rouja P, Roberts J, Seigel DA, Siuda A, Steinberg DK, Stevenson A, Sumaila VR, Swartz W, Trott TM,

Vats V. The protection and management of the Sargasso Sea: The golden floating rainforest of the Atlantic Ocean: Summary Science and Supporting Evidence Case Bermuda, BM. Bermuda: Sargasso Sea Alliance and Government of Bermuda. 2011. 44 p.

[94] Muñoz-Bautista AN. Composición taxonómica y abundancia de la macrofauna asociada a *Sargassum* (Phaephyceae: Fucales) flotantante en el Sistema Arrecifal Veracruzano, suroeste del Golfo de México [thesis]. Veracruz, México: Universidad Veracruzana; 2013.

[95] Fleury AG, Drazen JC. Abyssal scavenging communities attracted to *Sargassum* and fish in the Sargasso Sea. Deep Sea Research Part 1: Oceanographic Research Papers. 2013; 72: 141-147. DOI:10.1016/j.dsr.2012.11.004.

[96] Smetacek V, Zingone A. Green and golden seaweed tides on the rise. Nature. 2013;504:84-88. DOI:10.1038/nature12860

[97] Webster R K, Linton T. (2013). Development and implementation of *Sargassum* early advisory system (SEAS). Shore & Beach. 2013;81(3): 1-6. http://www.sargassoseacommission.org/storage/Webster_et_linon_2013_1.pdf

[98] Taylor, W.R. & Arndt, C.H. Marine algae of the southwestern Peninsula of Hispaniola. American Journal of Botany. 1929;16, 651-662.

[99] Tussenbroek BI, Hernández Arana HA, Rodríguez-Martínez RE, Espinoza-Avalos J, Canizales-Flores H, González-Godoy CE, Guadalupe Barba-Santos MG, Vega-Zepeda A, Collado-Vides L. Severe impacts of brown tides caused by *Sargassum* spp. on near-shore Caribbean seagrass communities. Marine Pollution Bulletin. 2017;122(1-2):272-281. DOI: 10.1016/j.marpolbul.2017.06.057

[100] Dreckmann KM, Sentíes A. Las arribazones de algas marinas en el caribe mexicano: evento biológico natural o basura en las playas. Biodiversitas CONABIO. 2013;107:7-11. ISSN: 1870-1760

[101] Laffoley D A et al. The Protection and Management of the Sargasso Sea: The Golden Floating Rainforest of the Atlantic Ocean 1-44 (Washington, 2011).

[102] Valiela I, McClelland J, Hauxwell J, Behr P, Hersh D, Kenneth F. Macroalgal blooms in shallow estuaries: Controls and ecophysiological and ecosystem consequences. Limnology and Oceanography 1997;42(5):1105-1118. DOI: 10.4319/lo.1997.42.5_part_2.1105

[103] Johnson DR, Ko DS, Franks JS, Moreno P, Sanchez-Rubio G. The *Sargassum* invasion of the Eastern Caribbean and dynamics of the Equatorial North Atlantic. In: Proceedings of 65th Gulf and Caribbean Fisheries Institute. Santa Marta, Colombia, November 5-9, 2012; pp. 102-103.

[104] Lapointe BE, West LE, Sutton TT, Hu C. Ryther revisited: nutrient excretions by fishes enhance productivity of pelagic *Sargassum* in the western North Atlantic Ocean. Journal of Experimental Marine Biology and Ecology. 2014;458:46-56. DOI: 10.1016/j.jembe.2014.05.002

[105] Rumphius, G.E. 1750. Herbarium amboinense ... pars sexta. Vol. 6. Amstelaedami [Amsterdam]; 1750. 256 p.

[106] Valmont de Bomare, J.C. Dictionnaire raisonné universal d'histoire naturelle. T. 3. Paris; 1764. 606 p.

[107] Lardizabal, V. Consuelo de Navegantes... Madrid: Oficina de D. Antonio Sanz, Impresor del Rey nuestro

Señor, y de su Real consejo; 1772. 230 p. https://play.google.com/books/reader?id=TRkKq1rC95AC&hl=es&pg=GBS.PT1

[108] Alquimar. Quienes somos. [Internet]. 2019. Available from: https://alquimar.com.mx/nosotros/ [Accessed: 2020-11-24]

[109] Desrochers A, Cox SA, Oxenford HA, van Tussenbroek B. *Sargassum* Uses Guide: A resource for Caribbean researchers, entrepreneurs and policy makers. Bridgetown, Barbados: CERMES, Food and Agriculture Organization of the United Nations (FAO), Technical Report No. 97; 2011. 172 p.

[110] Grupo MD. Productos. [Internet]. 2019. Available from: http://gpomd.com/dianco/nosotros [Accessed: 2020-11-24]

[111] Salgax Biotecnología Marina. Productos. [Internet]. 2019. Available from: http://www.salgax.com/productos [Accessed: 2020-11-24]

[112] Tapia-Tussell R, Avila-Arias J, Domínguez Maldonado J, Valero D, Olguin-Maciel E, Pérez-Brito D, Alzate-Gaviria L. Biological Pretreatment of Mexican Caribbean Macroalgae Consortiums Using Bm-2 Strain (*Trametes hirsuta*) and Its Enzymatic Broth to Improve Biomethane Potential. Energies. 2018;11: 494. DOI: 10.3390/en11030494

[113] Abaplas [Internet]. 2020. Available from: http://abaplas.com/ [Accessed: 2020-12-20]

[114] Arrellanos-Huerta ME. Desarrollo de un biopolímero a partir de residuos orgánicos [tesis]. Quintana Roo: Universidad Abierta y a Distancia de México; 2019.

[115] Renovare. Ova arena. [Internet]. 2020. Available from: https://www.renovareco.com/es/shop/ova-arena/?v=6f4f56f1b600 [Accessed: 2020-11-24]

[116] Dianco Mexico. [Internet]. 2020. Available from: http://gpomd.com/dianco [Accessed: 2020-12-20]

[117] Sarganico Productos Sustentables. El origen de sarganico. [Internet]. 2019. Available from: https://sarganico.mx/el-origen-de-sarganico/ [Accessed: 2020-11-24]

[118] Caamal-Fuentes E, Chale-Dzul J, Moo-Puc R, Freile-Pelegrin Y, Robledo D. Bioprospecting of brown seaweed (Ochrophyta) from the Yucatan Peninsula: cytotoxic, antiproliferative, and antiprotozoal activities. Journal of Applied Phycology. 2014;26:1009-1017. DOI: 10.1007/s10811-013-0129-x

[119] Chale-Dzul J, Freile-Pelegrin Y, Robledo D, Moo-Puc R. Protective effect of fucoidans from tropical seaweeds against oxidative stress in HepG2 cells. Journal of Applied Phycology. 2017;29:2229-2238. DOI: 10.1007/s10811-017-1194-3

[120] Chale-Dzul J, Pérez-Cabeza de Vaca R, Quintal-Novelo C, Olivera-Castillo L, Moo-Puc R. Hepatoprotective effect of a fucoidan extract from *Sargassum fluitans* Borgesen against CC4-induced toxicity in rats. International Journal of Biological Macromolecules. 2020;145:500-509. DOI: 10.1016/j.ijbio mac.2019.12.1830141-8130/

[121] Freile-Pelegrin Y, Robledo D, Chan-Bacab MJ, Ortega-Morales BO. Antileishmanial properties of tropical marine algae extracts. Fitoterapia. 2008;79:374-377. DOI: 10.1016/j.fitote.2008.02.006

[122] Aponte-Otaola NE, Diaz-Piferrer M, Graham HD. Seasonal Variations and Anatomical Distribution of Alginic Acid in *Sargassum* spp. Found

Along the Coasts of Puerto Rico. The Journal of Agriculture of the University of Puerto Rico. 1983;67(4):464-475.

[123] Mohammed A, Bissoon R, Bajnath E, Mohammed K, Lee T, Bissram M, Ward K. Multistage extraction and purification of waste *Sargassum natans* to produce sodium alginate: An optimization approach. Carbohydrate Polymers. 2018;198:109-118. https://www.sciencedirect.com/science/article/abs/pii/S0144861718307215

[124] Hernández-Carmona G. Variación estacional del contenido de alginatos de tres especies de feofitas de Baja California Sur, México. Investigaciones Marinas CICIMAR. 1985;2(1):29-45.

[125] Huang SX, Jiang Q, Ding YF, Wang FJ, Zhu C. Arsenic contents and speciation at different growth stages of *Sargassum fusiforme* (Harv.) Setchell (Hijiki), an edible seaweed. Applied Ecology and Environmental Research. 2020;18(1):1941-1952. DOI: 10.15666/aeer/1801_19411952

Chapter 5

The Endangered Species *Dioon edule* in the Sierra Madre Oriental in San Luis Potosí: Demography and Genetic Diversity

Gabriel Rubio-Méndez, Alberto Prado, Jacqueline C. Bede, José Arturo De-Nova, Joel Flores, Juan Antonio Reyes-Agüero and Laura Yáñez-Espinosa

Abstract

The distribution of the endangered species *Dioon edule* is in populations scattered throughout the Sierra Madre Oriental in San Luis Potosí, Mexico. Its habitat is tropical dry forests at lower elevations and oak forests at higher elevations, mainly disturbed by anthropic activities. We determined and analyzed nine populations' demographic structure and explored the genetic diversity of five using SSR markers. The population density averaged 2050 individuals ha^{-1} and have an aggregated distribution pattern. Differences in the plants' size among localities are due to site quality, based on their adaptation capacity and response to climate and soil traits. Most populations have the highest mortality in the early stages of life, with a low mortality rate for those who survive this bottleneck. Two populations have a relatively constant mortality rate, attributed to disturbance of the habitat. The populations show low genetic diversity and an excess of homozygotes. Their similarity is probably related to the formation of natural corridors favoring connectivity between populations. The deterioration and fragmentation of the habitat have severe effects on the populations' viability, like reducing gene flow, which has led to inbreeding and genetic drift.

Keywords: Zamiaceae, cycads, gymnosperms, ecotone, dioecious, habitat disturbance, inbreeding, Mexico, population, gene flow

1. Introduction

The relevance of the species *Dioon edule* (Zamiaceae) is evolutive because it belongs to a group of plants whose origin can be traced back to 250 million years ago, allowing us to understand the adaptations developed to survive the present day. It is also ecologically relevant as it is scattered in a wide range of ecosystems, from coastal dunes in the Gulf of Mexico's coastal plain to pine forest in the Sierra Madre Oriental, and its interactions with symbiotic micro-organisms, pollinating insects, and predator-dispersing wild and domestic fauna. The region's cultural relevance is defined by the Xi'iuy ethnic group in San Luis Potosí, collecting the seeds for food and the leaves for ceremonies [1].

This species is distributed in the Sierra Madre Oriental (SMO), within the geographic-cultural region known as La Huasteca, including the south of Tamaulipas, east of San Luis Potosi, Hidalgo, Queretaro, and Veracruz. The elevation goes from sea level to 1525 meters above sea level (m.a.s.l.) [2, 3]. It has stems up to 6 m high in some occasions semi-prostrated, presenting reproductive events with an interval of 10 to 52 years in female individuals and 2.8 to 8.8 in male individuals [4]. Its strobilus is pollinated only by specialist insects (beetles) associated with them [5].

The species is currently classified as near threatened (NT) by International Union for Conservation of Nature and endangered by the NOM-059-SEMARNAT-2010 in Mexico. This category is mainly because of land-use change to agriculture and livestock, causing habitat fragmentation that adversely affects population dynamics and a possible reduction in gene flow among populations [4, 6].

It is necessary to integrate information from different areas, mainly demography and population genetics, to develop and implement effective conservation strategies [6] to build general conclusions from the observations derived from each of these areas.

Demographic studies provide the basis for knowledge of populations and allow us to assess populations' viability, identify factors that reduce the population, and predict the persistence of small, isolated populations and evaluate alternative conservation scenarios [7]. Population genetics allows us to describe the genetic composition of natural populations and predict their changes in response to various evolutionary forces operating on them, allowing us to identify some risks of a genetic nature that affect the persistence of species such as habitat fragmentation and loss of adaptive potential associated with the decline of genetic diversity and inbreeding [8, 9]. Diverse studies related to population genetics and genomics of cycads have been published based on molecular markers such as Aloenzymes, RADseq, chloroplast markers, ISSR [4, 10–15].

The objective of this research was to study the demographic and genetic status of *Dioon edule* Lindl. (Zamiaceae) in representative natural populations of the Sierra Madre Oriental in the state of San Luis Potosi, analyzing their population structure concerning their density, age structure, sex ratio, as well as genetic diversity.

2. Materials and methods

2.1 Study area

The Sierra Madre Oriental is fundamentally a group of minor mountain ranges formed by marine sedimentary rocks, mainly limestone, sandstone, and shales. The maximum elevations are up to 3180 m.a.s.l., with small intermontane valleys limited by high mountain ranges with steep slopes (**Figure 1**). In this region, there are soils of alluvial origin formed in the great plains with sediments mostly coming from limestone and lutites and residual and colluvial origin in the high and lower parts of the mountain ranges constituted by the same type of rock. In general, the soil is medium texture, shallow, and rich in organic matter and nutrients depending on the climate and vegetation [16]. The area's climate is semi-warm, characterized by average annual temperatures above 18 °C and 600 mm of rainfall, mostly during the summer [16].

In the state of San Luis Potosi, three important vegetation regions for cycads stand out [16]: Tamasopo with tropical rain forest and oak forest (*Quercus* spp.); Aquismón with tropical rain forest; and, Rayón and Santa Catarina with oak forest, and piedmont scrub.

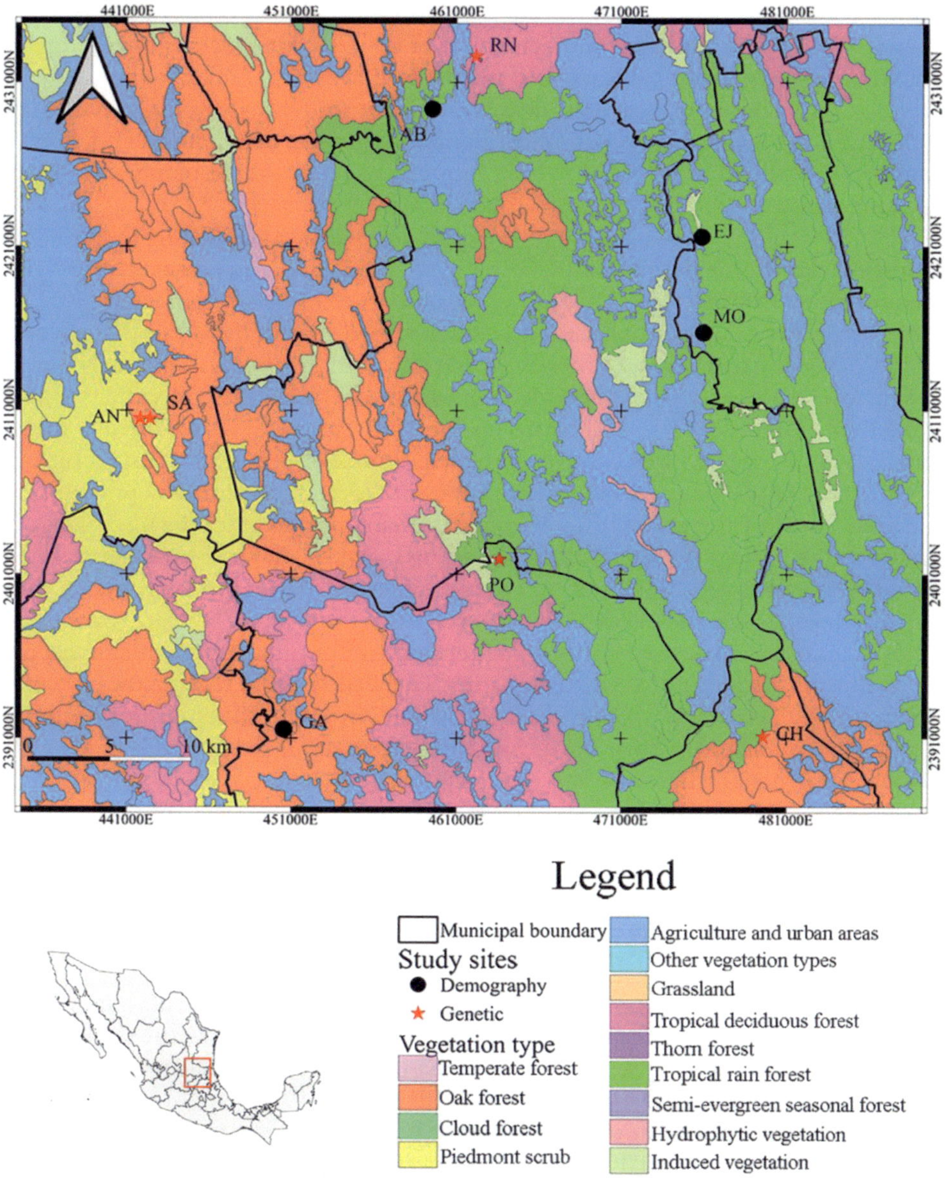

Figure 1.
Study area in the Sierra Madre oriental of San Luis Potosí, Mexico.

2.2 Demography

The study of demography was conducted in nine populations of *Dioon edule* distributed in the sub-province of Carso Huasteco (**Figure 1**) that covers 15.52% of the total surface of San Luis Potosi. It is formed almost exclusively by mountain ranges with soils of residual origin, shallow (less than 10 cm), with abundant rocky outcrops and Lithic phase. We do not provide detailed information on populations' locations to avoid possible looting, so each one has a key with two letters.

The sampling consists of modifying the line interception method [17] and used by [18] in transects of 30 m long and 10 m wide. Rectangular units provide the advantage of evaluating variables by walking in a straight line without moving sideways, starting at one end of the line, and considering only those plants within the plot [19]. Due to the harsh characteristics of the area with rocky slopes of up to 70°, we traced two linear transects 100 m long by 2 m wide, parallel to the slope, locating them where the most significant number of individuals were concentrated, considering that *D. edule* populations present an aggregate distribution [18]. Each transect was divided into four 25 m long subunits to carry out the plants' counts and measurements.

The different stages of the cycads' life cycle were defined according to the number of leaves, the number of leaflets and stem height [7] because it has not yet been possible to define the age of the cycads due to their growth characteristics [18]. Therefore, individuals were classified as follows: seedlings, individuals that present one or more leaves without these forming a crown; juveniles, individuals that present one or more crowns, but with a non-visible stem; adults, individuals that present one or more crowns and have a visible stem. Adults were classified as male, female, or non-reproductive according to the presence of reproductive structures.

Once the cycads' life cycle stages were defined, each stage's survival probabilities calculation as the proportion of individuals that survived to enter the next category, relative to the initial number of seeds ($l_x = n_x/n_0$) [20]. A survival curve describes the survival pattern of individuals in a population over time, plotting the lx on the y-axis, in logarithmic scale, with life cycle stages on the x-axis There are three basic survival curves, the survival curve type I represents a population in which most of the organisms die in the older age categories. The type II curve represents a population in which the mortality rate is almost constant, and type III represents a population in which most individuals die during the first stages of the life cycle, and very few reach the intermediate and late age categories [20].

The stem's diameter and the plant's total height were measured, the total number of leaves was counted, and the length of those on the last crown was measured.

Considering that the species *D. edule* is dioecious and produce only one cone per reproductive event, we evaluated the number of plants with male and female cones present as a measure of the potential reproductive capacity of the population per reproductive event [21].

The data on population density and potential reproductive capacity was analyzed using a one-way Analysis of Variance (ANOVA) to determine the statistical difference among the nine populations studied (localities) with two transects in each ($n = 18$).

Data on plant size and number of leaves of individuals in reproductive and non-reproductive states were analyzed through a nested ANOVA (random-effects model II) to determine the statistical difference among localities and categories of plants' reproductive states in all localities and each location. When statistically significant differences between means occurred, a Tukey multiple comparison test was applied ($p < 0.05$). All analyzes were performed with XLSTAT software (Addinsoft Inc., Paris, France).

2.3 Genetic diversity

The genetic diversity was studied in five populations (**Figure 1**), covering an altitudinal gradient from 388 to 1050 m.a.s.l. In each population, leaf tissue was collected from 21 individuals (seven from each age category). The DNA extraction was done with the DNeasy Plant minikit® Kit (Quiagen, Valencia, California, USA), following the kit's protocol and using approximately 100 mg of leaf tissue from which the cuticle was previously detached. A total of seven microsatellite markers designed by [22] (**Table 1**) labeled M-13 (−21pb) to obtain fluorescent products [23, 24], and PCR products were separated and genotyped with the LI-COR 4300 DNA analyzer

Locus	Primer sequences (5′–3′)	Repeat	N	Size
ED3	F: GCATGAGGAGCTTGTTCCGT	(CT)19	2	123–127
	R: CTGTGAACTCCTGAAAGCATC			
ED5	F: AGGCATAAATGGCTAAGCATAC	(AG)16	5	136–148
	R: GCATTTCTAGTGGACAAACCAG			
ED6	F: ATGCAGATGAAACACACCC	(TGG)8	2	239–242
	R: TCCTAACCATCCATCACTACC			
ED9	F: CCTTGTGTTACTTTGAGCACC	(CAT)9int	5	244–268
	R: CAACAATGTAAGTGATGATGCC			
CAP5	F: CACTACCACCCCTATACCAC	(CT)23	3	225–241
	R: GACTTGAGCTTGTCTTTGTTG			
TOM5	F: CGTTTCCATTGGAGAGACAAG	(TC)10	2	224–226
	R: CCATCCAAGTGAGTGATACAAG			
1660	F: GGTGCTGAAGAGGAAGAAGAA	(GAA)16	4	194–230

Table 1.
SSR primers used. N, number of alleles; size, allele size range (bp).

(LI-COR Biosciences, Lincoln, Nebraska, USA) in a 6.5% polyacrylamide gel (LI-COR Biosciences, Lincoln, Nebraska, USA), at a wavelength of 700 nm.

It was calculated the average number of alleles per locus, percentage of polymorphic locus, the expected and observed heterozygosity, the fixation index and it was estimated the deviation of the Hardy–Weinberg Equilibrium using the GenAlex software [17], for all the samples and grouping the individuals according to their life stage. An Analysis of Molecular Variance (AMOVA) was performed on genetic distances. Wright's F statistics (Fis, Fst, Fit) was calculated for the whole set of populations and each age category with the GenAlex software [25], and also the degree of genetic differentiation between pairs of populations (Fst) was estimated for all the samples and grouped by life stage.

3. Results

3.1 Demography

The density of the nine locations averaged 2050 individuals $ha^{-1} \pm 293.36$. Density ranges from 3775 individuals ha^{-1} in SA to 775 individuals ha^{-1} in MO. The populations with the highest density of individuals are the localities of SA and GA, and the lowest density is MO, although there are no differences between localities in the number of individuals ha^{-1} of the populations (ANOVA, $df = 8$; $F = 2.287$; $P > 0.05$).

The stages of the life cycle (**Table 2**) showed that seedlings were the most abundant, with an average of 49.27% ± 5.00, varying from the lowest in RN to the highest in SA. The proportion of young plants averaged 23.05% ± 2.62, with CH standing out with the lowest proportion and RN with the highest. The proportion of non-reproductive adults was 11.39% ± 3.12, being absent in SA and RN localities, unlike CH, where they represent a third of the population. Reproductive adults represented 16.29% ± 4.48, being scarce in SA and RN, but in the last, they are almost half of the population.

Location	Seedling (%)	Juvenile (%)	Adult	
			Non-reproductive (%)	Reproductive (%)
SA	70.20	26.49	0.00	3.31
GA	57.89	19.55	10.53	12.03
AN	50.93	28.70	16.67	3.70
CH	53.49	6.98	30.23	.30
PO	43.37	28.92	14.46	13.25
AB	46.55	25.86	8.62	18.97
EJ	35.56	22.22	15.56	26.67
RN	20.93	32.56	0.00	46.51
MO	64.52	16.13	6.45	12.90
Average	49.27	23.05	11.39	16.29

Table 2.
Frequency of plants per stage of their life cycle in the locations.

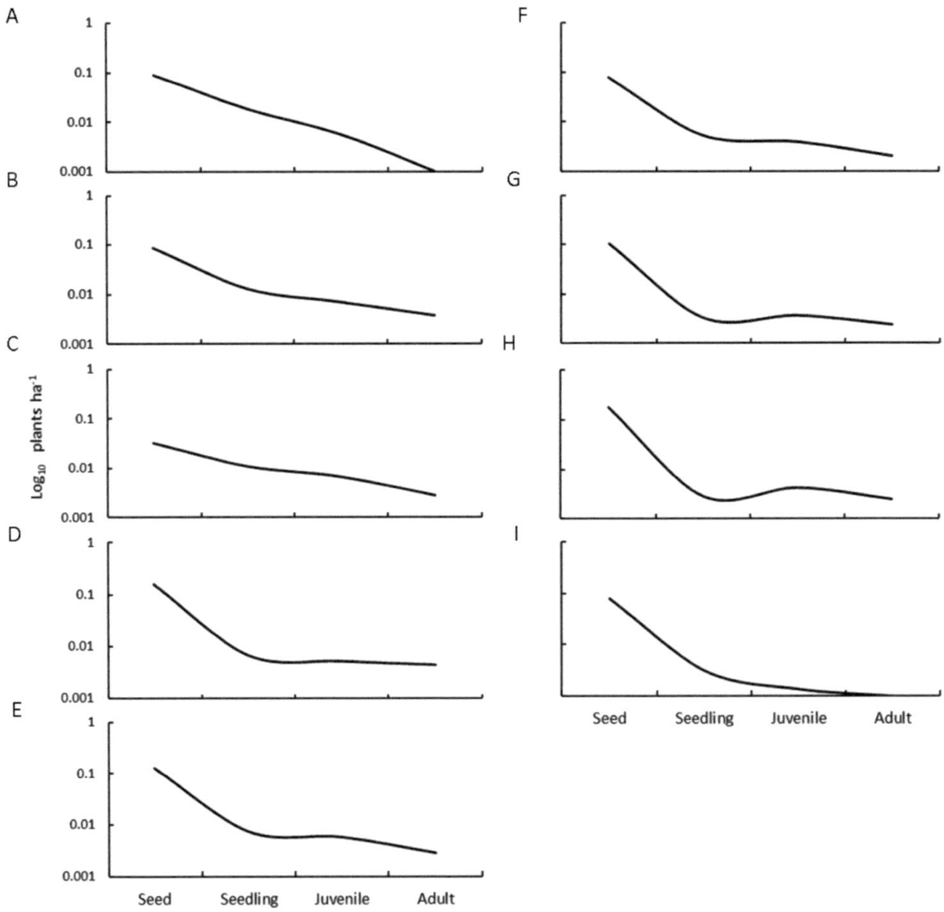

Figure 2.
Survival curves according to the life cycle stages of the nine populations of Dioon edule. *(A) SA, (B) GA, (C) AN, (D) CH, (E) PO, (F) AB, (G) EJ, (H) RN, (I) MO.*

The survival curve based on the states of the life cycle of the populations is an inverted "J" (type III), well defined in most locations (**Figure 2F, G, H, I**), and scarcely defined in other localities (**Figure 2D, E**). The localities of SA (**Figure 2A**), GA (**Figure 2B**) and AN (**Figure 2C**) showed a curve like type II in the populations.

The population's potential reproductive capacity was 250 ha^{-1} ± 45.26 individuals in the localities with reproductive structures, of which 158 ha^{-1} ± 35.36 were male and 92 ha^{-1} ± 17.18 female. The highest number of male individuals with the reproductive structure were GA and RN, and with the lowest number was SA (**Figure 3A**).

Likewise, the localities with the highest number of female individuals were RN and CH, and those with the lowest number were MO and AN (**Figure 3A**). The average sex ratio for the locations was 2: 1 (male: female), with the highest ratio of 4: 1 in GA and the lowest of 0.5: 1 in CH (**Figure 3B**). The ANOVA showed that there are no differences between mean localities in the number of female individuals ha^{-1} ($n = 18$; $df = 8$; $F = 1.234$; $P > 0.05$) and male individuals ha^{-1} ($n = 18$; $df = 8$; $F = 1.006$; $P > 0.05$) of the populations.

An average of 35.70% ±2.79 of individuals was found in the transect up to 25 m, mainly seedlings and juveniles in the localities. From 25 to 50 m was found on average 34.39% ± 5.60, in the 50 to 75 m 19.89% ± 3.41 and in the 75 to 100 m, 10.02% ± 2.44. The 70% of the individuals were concentrated in the first 50 m from the transect's starting point and particularly in RN 100%. At sites with a steeper slope, there were plants widely dispersed, while at sites with flatter slope (less than 20%, e.g. SA), there were large groups of individuals aggregated around adult plants or in the shade of oaks (*Quercus laeta*) and palms (*Brahea dulcis*).

Considering the plants' size and the number of leaves as parameters associated with the cycads' age, the longest and largest plants are in the communities of PO,

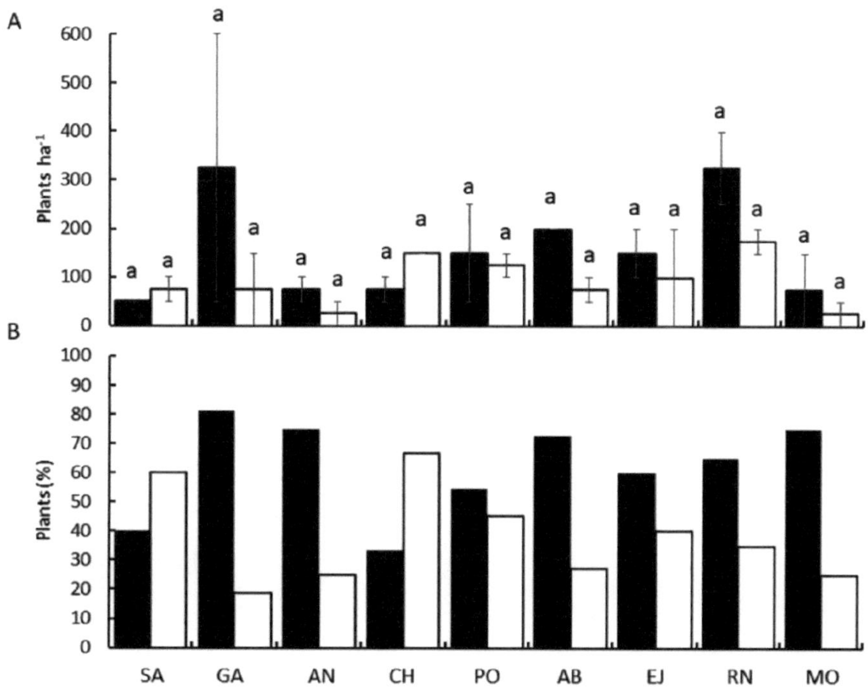

Figure 3.
Ratio of (A) the density of male (solid) and female (hollow) Dioon edule *plants (± SE) and (B) the sex ratio in each location. Different letters mean significant differences (Tukey <0.05).*

CH, GA, AN, and SA. If it was impossible to measure the adult plants' height and diameter because they had a subterranean stem, only the number of leaves was considered, obtaining similar results.

The average height for non-reproductive individuals was 23.4 cm ± 3.4, with a maximum average height of 52.2 cm and the lowest of 14.5 cm. For reproductive plants, the maximum recorded height was 56.3 cm, and the lowest was 9.5 cm, with an average of 28.8 cm ± 8.3. However, the ANOVA showed that there are differences among means in plant height between localities (df = 8; F = 5.428; P < 0.001) and between categories of reproductive status in localities (df = 8; F = 4.992; P < 0.05), but not between categories of reproductive status in each locality (df = 1; F = 2.633; P > 0.05) (**Figure 4**).

The average stem diameter was 12.2 cm ± 2.5 in non-breeding individuals, with a width of 23.6 cm to 15.0 cm (**Figure 4A**), and in reproductive individuals, the average was 16.5 cm ± 4.4 with a maximum of 31.2 cm and a minimum of 5.5 cm. The ANOVA showed that there are no differences in the mean diameter between localities (df = 8; F = 1627; P > 0.05) or between categories of reproductive status in each locality (df = 1; F = 2848; P > 0.05), but there are differences between the categories of reproductive status of individuals in the localities (df = 8; F = 4034; P < 0.05) (**Figure 4**).

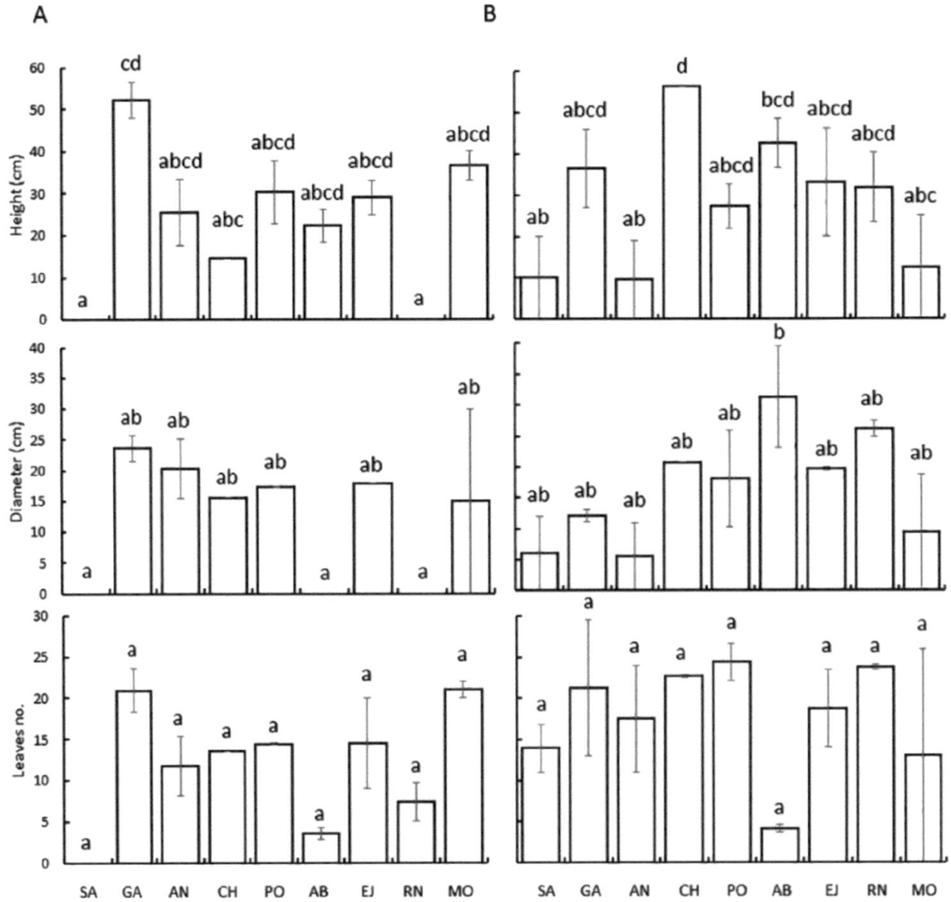

Figure 4.
Size of Dioon edule *plants (± SE) in each location. (A) Non reproductive and, (B) reproductive. Different letters mean significant differences (Tukey <0.05).*

The average number of leaves was 17.69 ± 2.18 in all individuals. The average number of leaves was 17.69 ± 2.18 in all individuals. In nonreproductive adults it was 14.3 with a maximum of 21 and a minimum of 3.5 (**Figure 4A**), for reproductive adults it was 15 with a maximum of 22.5 leaves. The ANOVA showed that there are differences in the mean number of leaves between localities (df = 8; F = 3.166; P < 0.05) and between categories of reproductive status in each locality (df = 1; F = 7.312; P < 0.05) but not between the categories of reproductive status of individuals in the localities (df = 8; F = 1.381; P > 0.05) (**Figure 4**).

3.2 Genetic diversity

Only three of the first seven proposed were amplified, ED9, TOM5, and CAP5. All the loci were polymorphic, but the population AN have the lowest proportion of polymorphic loci (33.33%) in the seedlings (**Table 3**). The average number of alleles per locus was 3.66; in the seedling stage, the lowest value corresponds to Los AN (1.33) (**Table 3**). There were no unique alleles within the adult stage. However, in the juvenile and seedling, four and three unique alleles were found, respectively, and ED9 was the locus that presented the highest number of unique alleles.

Location	Life cycle stage	N	Na	Polymorphic Loci (%)	Ho	He	F
SA	Seedling	6.33	2.67	100	0.111	0.464	0.795
	Juvenile	6.33	3	66.67	0.167	0.355	0.695
	Adult	6	2.67	100	0.133	0.489	0.810
	Total	18.67	3.67	100	0.137	0.496	0.817
AN	Seedling	3	1.33	33.33	0.000	0.163	1.000
	Juvenile	5	3.33	100	0.333	0.634	0.556
	Adult	6.67	3	100	0.143	0.511	0.791
	Total	14.67	4	100	0.194	0.583	0.741
CH	Seedling	6.67	3.33	100	0.167	0.613	0.786
	Juvenile	6.33	3.33	100	0.222	0.543	0.636
	Adult	7	2.67	100	0.048	0.541	0.934
	Total	20	4	100	0.139	0.600	0.811
PO	Seedling	7	3	66.67	0.333	0.429	0.239
	Juvenile	7	2.33	100	0.238	0.446	0.605
	Adult	7	2.67	100	0.238	0.418	0.605
	Total	21	3.33	100	0.270	0.446	0.572
RN	Seedling	6.67	2.67	100	0.325	0.334	0.197
	Juvenile	6.67	2.67	66.67	0.278	0.445	0.423
	Adult	7	2.33	66.67	0.286	0.303	0.354
	Total	20.33	3.33	100	0.297	0.394	0.260

Table 3.
Genetic diversity index for five Dioon edule *populations in San Luis Potosi, Mexico.*

Location	Locus	Seedling		Juvenile		Adult		Total	
		χ^2	P-value	χ^2	P-value	χ^2	P-value	χ^2	P-value
CH	CAP5	7.00	0.008	6.07	0.108	7.00	0.008	20.04	0.000
	ED9	11.33	0.332	14.04	0.171	15.52	0.017	38.85	0.001
	TOM5	14.00	0.003	7.00	0.008	7.00	0.008	42.00	0.000
RN	CAP5	7.00	0.008	14.00	0.003	7.00	0.008	42.00	0.000
	ED9	14.67	0.023	9.75	0.136	9.64	0.140	16.26	0.092
	TOM5	0.04	0.839	monomorphic		monomorphic		0.01	0.911
PO	CAP5	monomorphic		7.00	0.008	7.00	0.008	21.00	0.000
	ED9	5.07	0.535	2.29	0.514	7.88	0.247	10.25	0.115
	TOM5	21.00	0.002	7.00	0.008	7.00	0.008	63.00	0.000
AN	CAP5	monomorphic		4.00	0.046	6.00	0.014	11.00	0.001
	ED9	monomorphic		12.00	0.285	5.06	0.537	15.04	0.449
	TOM5	7.00	0.008	14.00	0.003	14.00	0.003	63.00	0.000
SA	CAP5	6.00	0.014	monomorphic		6.00	0.014	18.00	0.000
	ED9	3.34	0.343	24.67	0.055	8.89	0.180	31.50	0.008
	TOM5	14.00	0.003	7.00	0.008	7.00	0.008	42.00	0.000

Table 4.
Hardy–Weinberg equilibrium test for three microsatellite loci for five Dioon edule *populations in San Luis Potosi, Mexico.*

The average heterozygosity (Ho) was 0.207. The highest value was in the seedlings in PO with 0.333, and the lowest in the seedlings in PO (Ho = 0.0), and the adult stage in CH (0.048) (**Table 3**). For the expected Heterozygosity, the average was 0.504 and the highest value was 0.634 in the juvenile category for the PO locality.

In the populations, only two loci resulted in Hardy–Weinberg equilibrium (H-W), ED9 in the populations of RN, PO, and AN, and TOM5 in the population of RN (**Table 4**). In the seedlings, the ED9 locus resulted in H-W balance in the populations of CH, PO, and SA and TOM5 resulted in balance in RN. The loci CAP5 in PO and SA populations and ED9 in SA, were monomorphic, and CAP5 in PO and SA populations and ED9 in SA. In the juveniles, ED9 resulted in H-W equilibrium in all populations. However, in adults, this same locus resulted in equilibrium in most populations except CH (**Table 4**).

The AMOVA showed that the categories' differences are explained by the genetic variation between individuals for all populations and each age category (Seedlings = 50%; juveniles: 59%; adults: 62%: Total: 58%) (**Table 4**). The inbreeding coefficient (FIS) generally shows a deficiency of heterozygotes in all populations and age categories (Fis ≈ 0.666–0.708) (**Table 5**).

The fixation index (Fst) indicates a moderate differentiation in all populations, as well as in the age categories (Fst ≈ 0.11–0.115). The populations that share more alleles are CH and PO located in the south of the study area (Fst = 0.057; p = 0.006); those that share fewer alleles are PO and SA also in the south (Fst = 0.265; p = 0.001) (**Table 5**). The seedlings have a high genetic differentiation between pairs of populations ($Fst > 0.26$; $p < 0.05$), while in the juveniles, the differentiation is moderate ($Fst > 0.110$; $p < 0.05$), and in adults from moderate to high ($Fst > 1.15$; $p < 0.05$).

Life cycle stage	Source	df	SS	MS	VAR	%
Seedling	BP	4	20.63	5.16	0.27	26%
	BI	30	39.21	1.31	0.52	50%
	WI	35	9.00	0.26	0.26	24%
	Total	69	68.84		1.06	100%
Juvenile	BP	4	12.30	3.08	0.11	11%
	BI	30	44.93	1.50	0.60	59%
	WI	35	10.50	0.30	0.30	30%
	Total	69	67.73		1.01	100%
Adult	BP	4	11.53	2.88	0.10	11%
	BI	30	41.86	1.40	0.57	62%
	WI	35	8.50	0.24	0.24	26%
	Total	69	61.89		0.93	100%
Total	BP	4	31.21	7.80	0.15	15%
	BI	100	142.33	1.42	0.58	58%
	WI	105	28.00	0.27	0.27	27%
	Total	209	201.54		1	100%

Table 5.
Analysis of molecular variance (AMOVA) based on three microsatellite loci for five Dioon edule *populations in San Luis Potosi, Mexico. BP = between populations; BI = between individuals; WI = within individuals.*

4. Discussion

In the study area the average density was 2103 plants ha^{-1}, lower than the 3200–4633 plants ha^{-1} of *Dioon edule* in the center of Veracruz [4, 18], but higher than the 37 plants ha^{-1} of *D. purpusii* [26], 116 to 167 plants ha^{-1} of *D. merolae* [27] and 589 plants ha^{-1} of *D. holmgrenii* [28].

The absence of statistically significant differences between locations is mainly due to the parameters' wide variation observed. However, the populations with lower density coincide with other studies [27, 28], where the habitat's quality has decreased. The low quality of the habitat is due to the recurrent disturbance caused by agriculture, mainly the establishment of sugar cane fields in RN, pastures for livestock in EJ and MO, and the extraction of complete plants or their parts, like seeds and leaf harvest for tamale production [1, 29, 30]. In general, plant density and size are population attributes that indicate habitat quality [7].

Most populations presented the survival inverted "J" curve (type III) typical of shade-tolerant species [20] and most populations of *D. edule* [18], *D. purpusii* [26], and *D. angustifolium* [31]. These populations have a higher density of seedlings and juveniles with high mortality and few very long-lived adults with reduced mortality. Fewer populations have a relatively constant mortality rate throughout their life cycle, different from most *Dioon* populations. This modification could be attributed to the reduction of populations caused by habitat disturbing [32] since the natural forest was replaced with cattle pastures.

The plant's reproductive capacity is of great interest in conservation since it significantly impacts management and recovery strategies [33]. It is remarkable the high number of female plants present compared with other studies. The imbalance of females is probably related to climatic factors, soil quality, or the occurrence of fires, which in the study area is very common due to agricultural activities [18, 26, 34–38]. The plants' sex ratio average was 2: 1 (male: female) in all locations. However, the ideal ratio for *Dioon* species is 3:1 to ensure an abundant amount of pollen during female strobili receptivity [13]. There are five populations of concern that could present pollen adequacy problems during fertilization, mainly the CH locality that presented a ratio of 1:2.

The *Dioon edule* populations' distribution pattern is aggregated, which has long been attributed to the strobilus remaining attached to the plant's stem during dispersal, causing many seeds to remain close and around. Seed dispersal agents are currently being studied to identify them in the study area, including reptiles, birds, and mammals. It is relevant the evidence of asexual multiplication, which means that sprouts could be an effective strategy when sexual reproduction is difficult or in response to some stress. Like the cycad *Cycas armstrongii,* which produces sprouts after fires around the charred stems [36], facilitating rapid propagation and ensuring populations' maintenance [39].

The differences in height between reproductive and non-reproductive individuals have been attributed to the lowest growth rate of reproductive individuals due to the compression of stem tissues by the increase in weight of stored water and the production of heavy strobili [18, 40]. However, in this study, no significant differences were observed in any location, indicating that this difference does not always apply to all cycad populations.

Considering that there were significant plants' height differences among populations, it would be an indicator of the site quality. The sites of CH, GA and AB that presented the tallest plants would have the best quality since they present the lowest proportion of organic matter and smooth slopes. Diference in height and diameter of *Ceratozamia matudae* [41], *Dioon purpusii* [26] and *D. holmgrenii* [28] is attributed to disturbance conditions.

The plants' diameter did not present significant differences between localities. It has already been observed in *Cycas micronesica* [40] between reproductive and non-reproductive individuals, which can be attributed to the manoxylic wood type with a lot of succulent tissue [42]. In the same way, the difference in the number of leaves present in reproductive and non-reproductive individuals at each location and the difference between locations can be attributed that during the adult stage, the incidence of solar radiation correlates with leaf production, but the production of strobili in female individuals stops the production of new leaves, due to the greater allocation of resources for the development of the strabo [43]. The number of leaves in reproductive adults was lower than in *Dioon holmgrenii*, but much higher in non-reproductive ones [28].

As highlighted in this study, another determining factor in the development of *Dioon edule* populations is the deterioration of forests caused by agricultural activities [44]. These activities cause a lower rate of reproduction, capacity to adapt to changes, and less ability to establish themselves during the passage from one phenological stage to another, altering the age structure of the population [7].

The genetic diversity analysis tried to clear the picture of the medium and long-term repercussions of adverse factors identified when analyzing the species' population demography. However, we consider that the number of populations and molecular markers should be increased since *Dioon edule is ditributed* in various habitats with different microenvironmental conditions, causing phenotypical differences that should be analyzed [3, 31].

The populations analyzed show low genetic diversity and an excess of homozygotes in the seedlings, juveniles, and adults. Contrary to that expected for adult plants, which were believed to have become established before the disturbances. Similar behavior was present in a study conducted in Veracruz, mentioning that populations outside the Pleistocene flora and fauna refuges located in Veracruz and Oaxaca probably suffered a bottleneck due to exposure to adverse environmental conditions generating a significant decrease in population size reflected in the low levels of observed heterozygosity detected [10].

Previous studies suggest that life history issues, such as geographic distribution area, reproduction systems, and pollen and seed dispersal systems, have essential consequences on genetic structure and genetic variability levels in natural populations [4, 6, 10, 45, 46]. These factors are especially relevant in cycads because of their distribution in particular areas. They also present vegetative reproduction forming inside the populations [47].

In addition to the patches of clonal sprouts, the seedling banks that form around adult plants of *Dioon edule* and trees of *Quercus laeta* and *Bursera simaruba*, act as an extensive reservoir of genetic variation. However, this eventually is lost due to the high mortality rates that this species presents in the establishment's early stages. The mortality occurs naturally in populations because seedlings and juveniles slowly develop features that allow them to tolerate adverse conditions, such as prolonged drought or insolation [48]. These conditions can be aggravated by habitat fragmentation [6, 7, 44], causing gene drift in the population and making natural selection less effective by expressing deleterious or poorly tolerant genotypes to current environmental conditions [49].

Deviations in H-W were at most of the loci assessed, all tending towards a deficit of heterozygotes. The deviations may be due to the lag in the reproductive season, which can vary from four to 52 years in female plants [4]. This produces a time barrier that prevents individuals' panmictic crossing, causing groups to form that can function as independent populations.

Five unique alleles were identified with low frequencies in the ED9 and TOM5 locus in seedlings and juveniles. These alleles' appearance could be attributed to recurrent mutations or gene flow by pollen dispersal at great distances [50]. These alleles may benefit from genetic drift promoted by population size reduction.

AMOVA indicates a high differentiation in seedlings (FST = 0.26), probably due to the reduction in gene flow, which has led to inbreeding within the populations. Simultaneously, in adults and juveniles, the values are very similar and may be maintained by self-incompatibility that creates a dependence on pollinators to facilitate reproduction [46].

Some populations share some alleles, suggesting that they were connected by natural corridors some time ago, mainly in the vicinity of the Sierra Gorda in the northern of Querétaro and south of San Luis Potosí. However, in the current generations of seedlings, the Fst values indicate that the populations analyzed are becoming isolated. The isolation is probably caused by physical barriers associated with anthropic activities increasing in the area. This situation diminishes the habitat conditions and leads to higher mortality, affecting pollinators and favoring inbreeding and gene drift [6].

5. Conclusions

Some *Dioon edule* populations have declined substantially due to land-use change, primarily for sugar cane fields and cattle pastures, but the overall decline is relatively low. Populations genetic diversity suggests that at some time, they were connected by natural corridors. However, when combining the own high mortality of the species in the seedling and juvenile stages, the reduction and fragmentation of its habitat, and the decrease of pollinators, it favors endogamy and genetic drift, whose effects are more pronounced in small populations.

Acknowledgements

GRM thanks the support of CONACYT (Scholarship for Postgraduate Studies 290685) and the PMPCA-UASLP. The Cycad Society financed the research through the project "The chamal (*Dioon edule* Lindl.) in the state of San Luis Potosí" and CONACYT (104739), granted to LYE. IICA financed the project "The Chamal, a toxic but also edible cycad" granted to AP. We thank Raymundo Mora for his support in the field and Marc Hersh for his laboratory support.

Author details

Gabriel Rubio-Méndez[1], Alberto Prado[2], Jacqueline C. Bede[3], José Arturo De-Nova[4], Joel Flores[5], Juan Antonio Reyes-Agüero[4] and Laura Yáñez-Espinosa[4*]

1 Faculty of Natural Sciences, Autonomous University of Queretaro, Queretaro, Mexico

2 Juriquilla Unit, National School of Superior Studies, Queretaro, Mexico

3 Department of Plant Science, Macdonald Campus, McGill University, Sainte-Anne-de-Bellevue, Canada

4 Institute for Research on Desert Zones, Autonomous University of San Luis Potosi, San Luis Potosi, Mexico

5 Consortium for Research, Innovation and Development for Arid Zones, San Luis Potosi, Mexico

*Address all correspondence to: lyaneze@uaslp.mx

References

[1] Yáñez-Espinosa L. Chamal (*Dioon edule* Lindl.) in the State of San Luis Potosí, México. The Cycad Newsletter. 2009; 32:19-21.

[2] de Luca P, Sabato S, Vazquez-Torres M. Distribution and variation of *Dion edule* (Zamiaceae). Brittonia. 1982; 34 355-362. https://doi.org/10.2307/2806710

[3] Whitelock L. Variation in the mexican cycad *Dioon edule* (Zamiaceae). The Botanical Review. 2004;70 (2):240-49. doi:10.1663/0006-8101(2004)070[0240:VITMCD]2.0.CO;2.

[4] Octavio-Aguilar P. Demographics and population genetics for three locations in *Dioon edule* Lindl. (Zamiaceae) in the state of Veracruz [thesis]. Xalapa, Veracruz, Mexico: Instituto de Ecologia, A.C.;2009. [Spanish]

[5] Maldonado-Ruiz MF, Flores-Vázquez JC. Entomofauna asociada a *Dioon* sp. nov. y actividad de los insectos polinizadores en San Jerónimo Taviche, Oaxaca, México. In: Proceedings of Cycad 2008 The 8th International Conference on Cycad Biology; 13-15 January 2008; Panama City, Panama: Memoirs of The New York Botanical Garden, 106 ; 2012. p. 295-300 [Spanish]

[6] Cabrera-ToledoD,González-AstorgaJ, Vovides AP, Casas A, Vargas-Ponce O, Carrillo-Reyes P, Nolasco-Soto J, Vega E. Surviving background extinction: Inferences from historic and current dynamics in the contrasting population structures of two endemic Mexican cycads. Population Ecology. 2019:61:62-73. https://doi.org/10.1002/1438-390X.1008

[7] Lopez-Gallego C. Demographic variation in cycad populations inhabiting contrasting forest fragment. Biodiversity Conservation. 2008;17:1213-1225. https://doi.org/10.1007/s10531-007-9263-6

[8] **Godoy JA. Genetics, molecular markers, and species conservation.** *Ecosistemas.* **2009;18:23-33. [Spanish]**

[9] Griffiths A, Miller J, Suzuki D, Lewontin D, Gelbart W. Genetics. 7th ed. Madrid: McGraw-Hill-Interamericana; 2002. 849p. [Spanish]

[10] González-Astorga J, Vovides AP, Ferrer M, Iglesias C. Population Genetics of *Dioon edule* Lindl. (Zamiaceae, Cycadales): Biogeographical and Evolutionary Implications. Botanical Journal of the Linnean Society. 2003;80(3):457-467. https://doi.org10.1046/j.1095-8312.2003.00257.x

[11] Xie J, Jian S, Liu N. Genetic variation in the endemic plant *Cycas debaoensis* on the basis of ISSR analysis. Australian Journal of Botany. 2005;53:141-145. https://doi.org/10.1071/bt04068.

[12] Yang, Y, Li, Y, Li, L-F, Ge, X-J, Gong, X. Isolation and characterization of microsatellite markers for *Cycas debaoensis* YC Zhong et CJ Chen (Cycadaceae). Molecular Ecology Resources. 2008;8:913-915. https://doi.org/10.1111/j.1755-0998.2008.02114.x

[13] Cabrera-ToledoD,González-AstorgaJ, Vovides AP. Heterozygote excess in ancient populations of the critically endangered *Dioon caputoi* (Zamiaceae, Cycadales) from central Mexico. Botanical Journal of the Linnean Society. 2008;158:436-447. https://doi.org/10.1111/j.1095-8339.2008.00868.x

[14] Zhan QQ, Wang JF, Gong X, Peng H. Patterns of chloroplast DNA variation in *Cycas debaoensis* (Cycadaceae): conservation implications. Conservation Genetics.

2011;12:959-970. https://doi.org/10.1007/s10592-011-0198-9.

[15] Clugston JAR, Kenicer GJ, Milne R, Overcast I, Wilson TC, Nagalingum NS. RADseq as a valuable tool for plants with large genomes—A case study in cycads. Molecular Ecology Resources. 2019;19;1610-1622. DOI: 10.1111/1755-0998.13085.

[16] Instituto Nacional de Estadística, Geografía e Informatica [INEGI]. Geographical synthesis of the state of San Luis Potosi. México, D.F.:1985. [Spanish]

[17] Franco-López J, de la Cruz AG, Navarrete SN, Flores MG, Kato ME, Sánchez CS, Abarca ALG, Bedia, CM. Handbook of Ecology. 2nd ed. México: Trillas;2005. 266 p. [Spanish]

[18] Vovides AP. Spatial distribution, survival, and fecundity of *Dioon edule* (Zamiaceae) in a tropical deciduous forest in Veracruz, Mexico, with notes on its habitat. American Journal of Botany. 1990;77:1532-1543. https://doi.org/10.1002/j.1537-2197.1990.tb11394.x

[19] Matteucci S, Colma A. Methodology for the study of vegetation. Serie de biología, monografía no. 22. Washington, D. C.: Secretaría General Organización de los Estados Americanos. Programa Regional de Desarrollo Científico y Tecnológico. 1982. [Spanish]

[20] Carabias J, Meave JA, Valverde T, Cano-Santana Z. Ecology and environment in the 21st century. 1st ed. Mexico: Pearson Educación de México, S. A., 2009. 267 p. [Spanish]

[21] Tremblay R L. The effect of population structure, plant size, herbivory and reproductive potential on effective population size in the temperate epiphytic orchid, Sarcochilus australis. Cunninghamia. 2006;9:529-535.

[22] Moynihan J, Meerow A, Francisco-Ortega J. Isolation, characterization and cross-species amplification of microsatellite loci in the cycad genus *Dioon* (Zamiaceae). Potential utilization in population genetics studies of *Dioon edule*. Molecular Ecology Notes. 2007;7:72-74. doi:10.1111/j.1471-8286.2006.01531.x.

[23] Shuelke M. An economic method for the fluorescent labeling of PCR fragments. Nature Biotechnology. 2000;18:233-34. https://doi.org/10.1038/72708

[24] Blacket MJ, Robin C, Good RT, Lee SF, Miller D. Universal primers for fluorescent labelling of PCR fragments—an efficient and cost-effective approach to genotyping by fluorescence. Molecular Ecology Resources. 2012:12: 456-463.

[25] Peakall R, Smouse P. Genalex 6: Genetic Analysis in Excel. Population Genetic Software for Teaching and Research. Molecular Ecology Notes. 2006;6 (1):288-95. doi:10.1111/j.1471-8286.2005.01155.x.

[26] Yáñez-Espinosa L, Sosa-Sosa F. Population structure of *Dioon purpusii* Rose in Oaxaca, México. Neotropical Biology and Conservation. 2007;2:46-54.

[27] Lázaro-Zermeño JM, González-Espinosa M, Mendoza A, Martínez-RamosM,Quintana-AscencioPF. Individual growth, reproduction and population dynamics of *Dioon merolae* (Zamiaceae) under different leaf harvest histories in central Chiapas, México. Forest Ecology and Management. 2011:261:427-439. https://doi.org/10.1016/j.foreco.2010.10.028

[28] Velasco-García MV, Valdez-Hernández JI, Ramírez-Herrera C, Hernández-Hernández ML, López-Upton J, López-Mata L, López-Sánchez, H. Structure,

stage heterogeneity and spatial dispersion pattern of *Dioon holmgrenii* (Zamiaceae). Botanical Sciences. 2016;94:75-87. https://doi.org/10.17129/botsci.258 [Spanish]

[29] Rubio-Méndez G, Chávez-Acuña IJ, Yáñez-Espinosa L, Fortanelli-Martínez J. Effect of seed collection on the structure and dynamics of wild populations of *Dioon edule* (Zamiaceae). The Journal of the Torrey Botanical Society. (in press)

[30] Tristán-Martínez E, Fortanelli-Martínez J, Bonta M. Toxic Harvest: Chamal Cycad (*Dioon edule*) Food Culture in Xi'Iuy Indigenous Communities of San Luis Potosi, Mexico. Journal of Ethnobiology. 2020:4:519-534. https://doi.org/10.2993/0278-0771-40.4.519

[31] Rubio-Méndez G, Yáñez-Espinosa L, Salinas-Rodríguez MM, Hernández-Hernández KA, Balderas-González D. Demographics and population structure of *Dioon angustifolium* (Zamiaceae) in northeastern Mexico. Botanical Sciences. 2019;97(4):685-690. https://doi.org/10.17129/botsci.2262 [Spanish]

[32] Keppel G. Notes on the natural history of *Cycas seemannii* (Cycadaceae). South Pacific Journal of Natural Science. 2001;19:35-41. https://doi.org/10.1071/SP01007

[33] Gaol ML, Fox ED. Reproductive potential of Acacia species in the central wheatbelt: variation between years. Conservation Science Western Australia Journal. 2002;4:147-152.

[34] Alejandre-Rosas J, Sanchez-Tinoco MY, Vazquez-Torres M. Population structure of *Ceratozamia mexicana* Brong. (Zamiaceae) in a forest in central Veracruz. La Ciencia y el Hombre. 1990;1:93-112. [Spanish]

[35] Ornduff R. Size classes, reproductive behavior, and insect associates of *Cycas media* (Cycadaceae) in Australia. Botanical Gazette. 1991;152:203-107. https://doi.org/10.1086/337880

[36] Watkinson AR, Powell JC. The life history and population structure of *Cycas armstrongii* in monsoonal northern Australia. Oecologia. 1997;111:341-349. https://doi.org/10.1007/s004420050244

[37] Nicolalde-Morejón, F. Ecology and Taxonomy of *Zamia* in Ecuador. Quito, Ecuador: Fundación Ecuatoriana para la Investigación y Desarrollo de la Botánica, Boletín 9; 2001. [Spanish]

[38] Pérez-Farrera MA, Vovides AP. Spatial distribution, population structure, and fecundity of *Ceratozamia matudae* (Zamiaceae) in El Triunfo Biosphere Reserve, Chiapas, México. The Botanical Review. 2004;70:299-311. https://doi.org/10.1663/0006-8101(2004)070[0299:sdpsaf]2.0.co;2

[39] Brenes-Cambronero L, Di Stefano JF. Possible influence of ramets on the population structure and distribution of the giant tree *Warszewicsia uxpanapensis*, cordillera Tilarán, Costa Rica. Revista Biología Tropical. 2006;54:1179-1188. https://doi.org/10.15517/rbt.v54i4.3095 [Spanish]

[40] Niklas K, Marler TE. Sex population differences in the allometry of an endangered cycad species *Cycas micronesica* (Cycadales). International Journal of Plant Sciences. 2008;169: 659-665. https://doi.org/10.1086/533606.

[41] Pérez-Farrera MA, Quintana-Ascencio PF, Salvatierra-Izaba B, Vovides AP. Population dynamic of *Ceratozamia matudai* in el Triunfo Biosphere Reserve, Chiapas, México. Journal of the Torrey Botanical Society. 2000;127:291-299. https://doi.org/10.2307/3088647

[42] Stevenson DW. Radial growth in the Cycadales. American Journal of Botany. 1980;67:465-475. https://doi.org/10.1002/j.1537-2197.1980.tb07674.x

[43] Clark D, Clark DB. Leaf production and the cost of reproduction in the Neotropical rain forest cycad *Zamia skinneri*. Journal of Ecology. 1987;76:1153-1163. https://doi.org/10.2307/2260640

[44] Donaldson, JS, editor. 2003. Cycads status, survey and conservation action plan. Cambridge, UK: IUCN/SSC Cycad Specialist Group. IUCN. 2003. ix+86 pp.

[45] Bohonak A. Dispersal, gene flow and population structure. The Quarterly Review of Biology. 1999;74(1):21-45. https://doi.org/10.1086/392950

[46] Mathiasen P, Rovere A, Premoli A. Genetic Structure and Early Effects of Inbreeding in Fragmented Temperate Forest of a Self-Incompatible Tree, *Embothrium coccineum*. Conservation Biology. 2006;21 (1):232-40. doi:10.1111/j.1523-1739.2006.00565.x.

[47] Cabrera-Toledo D, González-Astorga J, Flores-Vázquez J. Fine-scale spatial genetic structure in two mexican cycad species *Dioon caputoi* and *Dioon merolae* (Zamiaceae, Cycadales): Implications for conservation. Biochemical Systematics and Ecology. 2012;40:6. https://doi.org/10.1016/j.bse.2011.09.004

[48] Augspurger CK. Seedling survival of tropical tree species: Interactions of dispersal distance, light gaps, and pathogens. Ecology. 1984;65:1705-1712. https://doi.org/10.2307/1937766

[49] Allendorf FW, Luikart GH. Conservation and the Genetics of Populations. 2nd ed. UK:Blackwell Publishing. 2007 ISBN: 978-0-470-67146-7

[50] Acuña W, Yalta C, Veli E. Cross species transferability of microsatellite markers from *Anas platyrhynchos* to Peruvian Muscovy Duck *Cairina moschata domestica*. Revista Peruana de Biología. 2020:27:255-260. http://dx.doi.org/10.15381/rpb.v27i2.15015

Chapter 6

Rediscovering Kemp's Ridley Sea Turtle (*Lepidochelys kempii*): Molecular Analysis and Threats

Miguel Angel Reyes-López, Fátima Yedith Camacho-Sánchez, Catherine E. Hart, Valeria Leal-Sepúlveda, Kevin Alan Zavala-Félix, César Paúl Ley-Quiñónez, A. Alonso Aguirre and Alan Alfredo Zavala-Norzagaray

Abstract

Sea turtles are reptiles that have inhabited the earth for 100 million years. These are divided into 2 families (Cheloniidae and Dermochelyidae) and 7 species of sea turtles in the world: the leatherback turtle (*Dermochelys coriacea*); hawksbill turtle (*Eretmochelys imbricata*); Kemp's ridley (*Lepidochelys kempii*); olive ridley (*L. olivacea*); Loggerhead turtle (*Caretta caretta*); flatback sea turtle (*Natator depressus*) and green turtle (*Chelonia mydas*). In particular, Kemp's ridley is included in the red list of IUCN categorized as "critically endangered". The most important site around the Word is in Rancho Nuevo, Tamaulipas, Mexico. Where 80–95% of the world's nesting is concentrated. Other nesting areas are Tepeguajes and Barra del Tordo, in Tamaulipas, and with less intensity in Veracruz (Lechuguillas and El Raudal beaches) and South Padre Island, Texas, USA. They deposit an average of about 90 eggs and hatching takes 40 to 60 days. Therefore, they are vulnerable to different anthropogenic activities and sources of pollution, such as heavy metals, which can cause toxic effects that are harmful to the turtles, damage their physiology and health. To understand the real situation about health and genetic parameters it is necessary to analyze biochemical and molecular factors in this species.

Keywords: Kemp's ridley, molecular analysis, nesting beaches, pollution

1. Introduction

For decades, the Kemp's ridley (*Lepidochelys kempii*) was one of the most elusive sea turtles, and its nesting sites were unknown. Study, ingenuity and a home movie shed light on what for many was a riddle. The first references occurred when fishers off the coast of Florida, USA reported that they caught "demonic turtles". The turtle's morphology was described as flat and grey with a large head and that they were very active when they were caught in fishing nets which resulted in broken nets. The encounters were reported to researchers, and Samuel Garman, a prominent herpetologist and ichthyologist at the Harvard Museum of Comparative Zoology, was the first person to describe the turtle characteristics and, it is thanks to him that the Kemps ridley sea turtle received its name. In 1880, Garman named the species after Richard Kemp who had a fascination for natural history and sent specimens of the turtles to Garman for study. It was Kemp who first thought the turtles were hybrids from a mating between green (*Chelonia mydas*) and loggerhead (*Caretta caretta*) turtles, hence their nickname of bastard turtles [1].

Another authority in the kemps ridley story was Archie Carr, who wrote several publications on the species. Despite this, he was unable to locate the turtles nesting sites. However, one of his disciples and collaborators, Peter Pritchard, continued this work and established the species morphometric measurements and other key information, such as foraging sites, but the search for the nesting sites of the mysterious Kemp's ridley was unsuccessful.

1.1 Discovery

It took two further researchers and several years to start to unravel the mystery. One crucial player in the discovery was Ing. Andrés Herrera, a rancher and lover of the outdoors who lived in Tamaulipas, Mexico. Ing. Herrera listened to locals talk of beaches where sea turtles nested along the coasts of Tamaulipas. Luckily Andres was also a light aircraft pilot, armed with information that turtles nested on specific beaches in Tamaulipas during the spring, and he decided to start flying his aeroplane at that time to try and observe them. According to Ing. Herrera, on 18th June 1947, one of these flights was successful when he encountered a mass nesting event, or "*arribada*", which occurred on the beaches of the town of Rancho Nuevo near Barra Calabazas, Tamaulipas. He landed the aeroplane and made the famous film, "the film of Ing. Herrera". This footage not only documented Kemp's ridley nesting but also that this was a mass nesting event or "arribada" which was happening in broad daylight which was different to other sea turtles such as the green turtles which nest at night [2].

Another key to solving the kemps ridley riddle was the biologist Henry Hildebrand from the University of Corpus Christi who had heard and read about the stories of the *arribada* and the place where said nesting had taken place, as well as the writings of Archie Carr. At the beginning of the 1960s, Hildebrand learned footage made by Ing. Herrera and travelled to Tamaulipas to ask for permission to present it at the 1961 meeting of the American Society of Ichthyologists and Herpetologists at the University of Texas. After several decades of uncertainty, Hildebrand in 1963 presented the mystery of where this elusive turtle nested to the scientific community. Later, Carr and Hildebrand's reports estimated that the film's famous *arribada* accounted for around 40,000 Kemp's ridley nesters in a single day [3, 4].

1.2 Population

The Kemp's ridley turtle population baselines is an estimated 40,000 nesting females in a single day of an arribada. However, in 1966, the largest arribada was reported to include just 1,317 turtles [5] and the number of nests laid continued to decrease until the 1970s, when records show that in 1978, the number of nests fell below 1,000 [6].

Later, in the mid-1980s, only 702 nests were recorded at Rancho Nuevo, Tamaulipas. However, from 1985 to 1990, nesting stabilised, with nest numbers ranging from 702 to 839. Finally, the USA and Mexico introduced legislation to reduce bycatch, protect major nesting sites, and establish a binational agreement. These crucial changes contributed to the increase and later recovery in the numbers of nesting turtles, which increased from approximately 1,000 nests in the mid-1990s to over 21,000 nests on the beaches of Tamaulipas in 2009 [7–9].

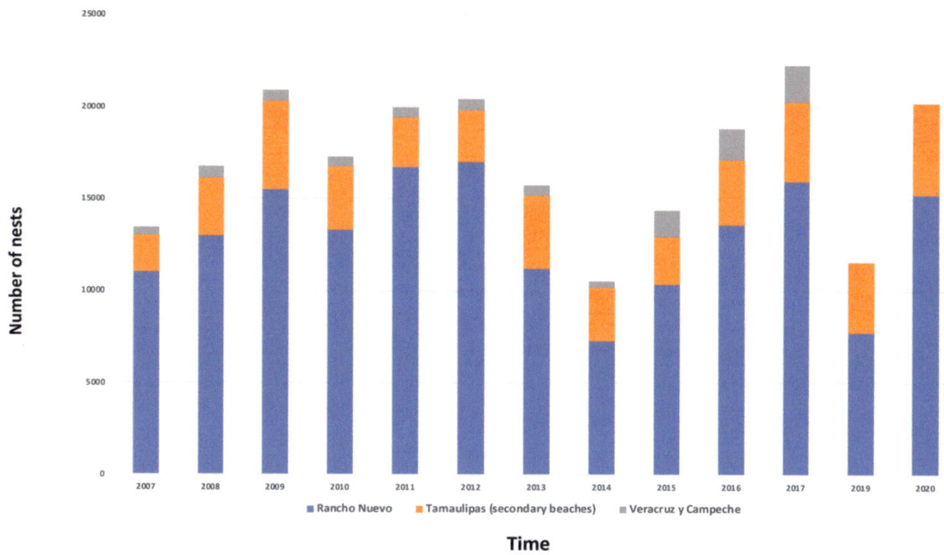

Figure 1.
Number of protected Kemp's ridley (Lepidochelys kempii) nests in Mexico from 2007 to 2020 [10].

By the 2011–2012 nesting season, the kemps ridley turtle population had stabilised at 21,000 nests. However, in 2013 numbers of nesting turtles fell rapidly to around 16,000, and the downward trend continued with just 12,000 nests recorded in 2014. Over the last five years (2015–2020), an average of 1,500 nests have been protected per season at the Rancho Nuevo Sanctuary compared to the 3,868 on the other Tamaulipas beaches. During the 2020 nesting season, there were encouraging results with 15,210 nests in Rancho Nuevo and 4,995 from other Tamaulipas beaches (**Figure 1**). It is important to note that data for 2018 was unavailable and the data for 2019 and 2020 do not include the number of nests for beaches found outside of Tamaulipas state [10].

2. Biology (lifecycle and reproduction)

2.1 Morphology

The Kemp's ridley sea turtle (*Lepidochelys kempii*) is one of the seven species of sea turtles that exist worldwide. The species belongs to the Cryptodira turtles of the Cheloniidae family and is a direct relative of the olive ridley sea turtle *Lepidochelys olivacea* (**Table 1**). Studies on Kemp's ridley sea turtle genetics show that there is only one population within this species, which evolved and has existed for approximately 2.5–3.5 million years [11]. The Kemp's ridley is the smallest sea turtle and is an endemic species to the Gulf of Mexico.

The Kemp's ridley (**Figure 2**) has usually a carapace as wide as it is long and contains five pairs of costal scutes that overlap the bony carapace (**Figure 2A, B**) and a triangular head with a slightly hooked bill (**Figure 2C-E**). The hatchlings are dark in colour on both sides. Adults generally present a greyish-green carapace with a pale yellowish plastron Each of the front flippers has a claw, while the rear flippers can present one or two claws. Juveniles have a dark grey carapace with a yellowish-white plastron, and adults develop an olive-grey or dark green carapace and a cream or yellowish plastron [12].

2.2 Taxonomy and nomenclature

Domain	Eukaryota
Kingdom	Animalia
Phylum	Chordata
Class	Reptilia
Order	Testudines
Family	Cheloniidae
Genus	*Lepidochelys*
Species	*kempii*

Table 1.
Kemp's ridley sea turtle (Lepidochelys kempii) taxonomy [13].

Figure 2.
Pictures showing general morphology of Kemp's ridley. (A and B) general view; (C) Aeria l view; (D) Lateral view; and (E) Frontal view. Pictures courtesy: Fátima Y. Camacho-Sánchez.

2.3 Life cycle

The Kemp's ridley sea turtle presents characteristics that differentiate it from other species, in addition to its small size and restricted distribution, this species, nests during the day, and as with the olive ridley sea turtle nests on mass in a phenomenon known as an "arribada". Kemps ridley females deposit an average of 90–100 eggs per nest. These turtles nest 2 to 3 times per season, with an interval of 14–28 days between laying each clutch of eggs. Another distinctive feature are the pores located in their plastron, known as "Rathke's gland", through which they secrete a substance that is considered a pheromone [14].

2.4 Biogeography

Globally the Kemp's ridleys sea turtle is represented by one unique population which presents one of the most restricted distributions of all sea turtle species. Kemp's ridleys are found primarily in the Gulf of Mexico and secondarily in the Atlantic Ocean, on the south-eastern coasts of the United States (USA). Approximately 70% of the population nests on a beach with a linear extension of 30.4 km, located in Rancho Nuevo, in Tamaulipas, Mexico. Ninety-seven per cent of the population lives in an area of 146 km that includes Rancho Nuevo beach and surrounding areas. A small proportion of the population is found on the coast of Veracruz, Mexico and the coast of Texas, USA (**Figure 1**). In southern Texas, South Padre Island is considered a secondary rookery for Kemp's ridley nesting after Rancho Nuevo [15].

On rare occasions, they are found off the coasts of Canada, Bermuda, Azores, Madeira and in the Mediterranean Sea [16]. However, the Gulf of Mexico is more than just the nesting ground for Kemp's ridley turtles. In addition to navigating neritic areas, both hatchling and juvenile Kemp's ridley turtles use oceanic areas of the Gulf of Mexico during their development, by entering the open ocean to forage. During migration, turtles can be found in the Northwest Atlantic and the Mediterranean Sea (**Figure 2**), for example, Bermuda, Canada, France, Ireland, Portugal and the United Kingdom [13, 16].

The Kemp's ridley turtle shares many characteristics with other sea turtle species in terms of its life cycle. Female turtles nest on beaches, depositing their eggs beneath the sand. The eggs hatch after approximately 45–58 days and the hatchlings migrate to the sea. The following years are spent on foraging grounds where they develop into adults at which point, they return to their natal area to reproduce and nest. During the sea turtle lifecycles, three key phases occur, and these are divided by ecosystem: terrestrial phase, neritic phase, and oceanic phase.

2.5 Terrestrial zone (Hatching)

The Kemps ridley turtle uses terrestrial ecosystem during three stages of its life cycle: nesting females, eggs, and hatchlings. The principal Kemp's ridley nesting beach at Rancho Nuevo is characterised by small dunes of variable size, with vegetation that improves stability. The sand is small and fine-grained, and the beach is high energy. The nesting season runs from March to August [15]. Mass nesting events or *arribadas* occur mainly in early April through the July period. These phenomena can be triggered by meteorological conditions represented by strong winds, primarily from the north, or by a change in atmospheric pressure [17].

Kemp's ridley nesting typically occurs during the day; however, reports of sporadic cases of nocturnal nesting exist in Texas. Clutch incubation lasts between 45–60 days, and climatic conditions act directly on embryo development with incubation duration and hatchling sex ratios affected by temperature. Studies have shown that sex ratios are male-biased at the beginning of the nesting season when temperatures are lower. In contrast, nests produce mainly female hatchlings when laid during the second arribada when temperatures are highest [18].

2.6 Neritic zone (juvenile stage and adult stage)

The hatchlings, after being released on the sand, immediately begin their trajectory to the ocean. The hatchlings swim using the energy provided by the nutrients found in the yolk of their egg, which lasts for a maximum of 4 days, during this period, the hatchlings migrate to ideal areas for their development. Hatchlings enter the surf and orient seaward into the waves, diving as they break, thus being swept seaward by the motion of the waves without significant energy expenditure. Like other species, the hatchlings can perceive the waves' movements to guide them towards the sea, the same mechanism they use on the way back to the coast, recognising the magnetic orientation. Juveniles reach adulthood in neritic waters, which are their most frequently used habitat. The Gulf of Mexico's northern coast is home to the majority of kemps ridley sea turtles, principally in waters no greater than 37 m deep. Although kemps ridley turtles may consume algae, their diet is primarily carnivorous and primarily based on crabs, but occasionally includes clams, shrimp, jellyfish and some fish species. Turtles may also scavenge in benthic areas on marine debris [15].

2.7 Oceanic zone

The hatchlings that manage to reach oceanic or pelagic areas stop actively swimming and enter a passive state, allowing themselves to be guided by the ocean currents to foraging grounds where they remain as they develop into juveniles. Although kemps ridley sea turtles are considered a carnivorous species, individuals have been seen feeding on sargassum [15]. During their oceanic phase, juvenile turtles can be divided into two groups: the current system of the northern and western Gulf of Mexico or the Gulf Stream of the Northwest Atlantic. However, a small proportion of the population finds its way to the Mediterranean Sea. During this period that lasts between 1 to 4 years, juveniles complete their oceanic feeding phase to return to surface waters within the Gulf of Mexico and the northwest Atlantic, in US waters, where they continue to forage.

3. Status

3.1 Threats

Sea turtle populations face multiple threats which need research and monitoring to understand how these threats are detrimental to their populations and conservation. Overlooking the multiple threats that species face throughout their life cycle puts the success of recovery plans at risk [19].

3.2 Illegal hunting: direct consumption and illegal trade

Globally, fishing and eggs collection for human consumption are the principal causes of the drastic reduction in sea turtle populations, and the Kemp's ridley is no exception. In the late 1960s, excessive capture of adult kemps ridley turtles and their eggs, contributed to the drastic reduction by almost 99% of the Kemp's ridley nesting population. Turtle eggs are considered an aphrodisiac in some countries [19]. The consumption of turtle eggs is not as frequent in the United States of America and Mexico as it was before the federal government bans on harvest and consumption. However, despite governmental agencies and non-governmental organisations efforts with protection and public awareness programs, consumption continues in some coastal communities.

3.3 Climate change

Climate change can have a severe impact on turtle populations. All reptiles, including sea turtles, are dependent on environmental temperature, which they use to regulate their physiology [20]. Sex determination in sea turtles is directly dependent on environmental temperature. Therefore, projected anthropogenic climate change will alter the sand temperature, affecting the primary sex ratios and increasing the risk of instability in the composition of sea turtle populations.

3.4 Marine pollution: rubbish and oil

Sea turtles can mistake plastic objects floating in the water column for food, such as jellyfish. If ingested these inorganic materials can choke or cause obstructions in the turtle digestive systems. They can also become entangled in discarded fishing lines and nets, drowning or unable to feed or swim. Trash on beaches can trap hatchlings and prevent them from reaching the sea.

During 2010 an accident occurred on the Deepwater Horizon platform that triggered an oil spill from the platform into the Gulf of Mexico This event negatively impacted the marine ecosystem, including the kemps ridley turtle as their essential habitats and migratory routes were affected by the spill [21]. However, the extent to which this oil spill was responsible for the decrease in the Kemp's ridley population observed in the years following the disaster is unknown, with other factors possibly contributing to the decline. Continued monitoring is essential to understanding the long-term effects of this and other oil spill events.

3.5 Bycatch

Every year thousands of turtles are accidentally caught by the shrimp fishery. Sea turtles must surface to breathe, so prolonged periods trapped in submerged fishing gear leads to drowning. Fisheries that use longlines and gillnets are also major causes of sea turtle mortality [22]. Although laws requiring the use of sea turtle excluder devices in shrimp fishery exist, turtles continue to become trapped and drown in these nets.

3.6 Coastal development: beach modification and human presence

Uncontrolled coastal development has destroyed beaches that are essential for nesting. The lights coming from roads and buildings confuse hatchlings and disorienting them away from the sea. Vehicles used in beach restoration projects, including dredging and sand nourishment projects, damage near-shore foraging areas and beaches, and can also destroy nests and hatchlings. These activities referred to as "beach maintenance" have been reported on Padre Island, where machinery has prevented hatchlings from reaching the water, leaving them trapped and exposed to vehicle traffic [15].

3.7 Marine biotoxins

Harmful algal blooms in the oceans are natural phenomena that occur as a result of increasing temperature, alteration of ocean currents, intensification or weakening of local nutrient upwelling, and heavy precipitation and storm events causing changes in land runoff [23]. Brevetoxins are a group of biotoxins produced mainly by algae *Karenia brevis*, which is the cause of the main harmful algal blooms (HABs) along the coast of south-west Florida, with periodic blooms throughout the Gulf of Mexico [24].

Sea turtles are affected by marine biotoxins present in the water column, in aerosols generated by waves, or through the consumption of contaminated prey [25]. In recent years, the impacts of these phenomena have caused or contributed to sea turtle mass mortality events in *C. mydas*, *C. caretta*, *L. olivacea* and *L. kempii* [26].

4. Studies

4.1 Satellite telemetry

Studies that follow in water movements of sea turtles and other marine vertebrates often use tagging to shed light on migration routes and local movements. The most common ways to tag sea turtles are to apply external metallic tags with unique codes usually to the turtle flippers which allow the individual turtle to be identified. This method requires the turtle to be recaptured and does not supply information on the routes taken from release and capture position. The advent of satellite telemetry in the 90s and improving technology has allowed researchers to follow the routes taken by sea turtles and better understand habitat use and connectivity [27].

Satellite tagging of Kemp's ridley sea turtles, as with other sea turtle species requires the attachment of platform terminal transmitters PTT to the turtle's carapace using epoxy glue. The study of the migration of kemps ridley turtles began in 1995 and has focused mainly on post nesting females. These marked females confirmed that Kemp's ridley turtles keep mainly to the Gulf of Mexico, and forage near the North Texas and Louisiana coasts in the USA. Satellite tracking of juvenile turtles shows that they live off the coast of Veracruz and that they nest mainly in Tamaulipas and more specifically in Rancho Nuevo Sanctuary. Additionally, information on foraging site fidelity is easily obtained from satellite telemetry studies. Furthermore, studies have also indicated that male Kemp's ridley turtles typically remain off the coast of Tamaulipas, close to the principal nesting area [28].

Satellite telemetry has also shed light on nesting and nest-site fidelity. Tracked females have been shown to nest one to three times at the same site. Additionally, the distance females migrate from nesting sites to the open ocean decreased as the number of times they returned to nest increased. Somewhat surprisingly satellite telemetry has shown that turtles do not forage during the inter-nesting period, and they minimise energy loss by spending most of their time resting, limiting movement.

Of the studies on female kemps ridley movements between nesting events, information is only available for three individuals. These studies found that after nesting one tagged female travelled north [29] from the main nesting site at Rancho Nuevo Sanctuary; two turtles have travelled south from Rancho Nuevo with one covering a distance of ~100 km Further studies are required to understand Kemps ridley movements between nesting events.

4.2 Biochemical studies

Sea turtles play an important role in ocean ecosystems, by maintaining healthy seagrass beds, beaches, estuaries, and reefs. In turn, they are considered sentinel species for ecosystem health, since their longevity and physiology provide essential and early information on marine and coastal habitats and the local environments in which they live, which provides a quick risk diagnosis. These reptiles are particularly susceptible to pathogens such as parasites, bacteria, fungi and viruses. Examples of these parasites are for bacteria: *Vibrio* spp. *Pseudomonas* spp., *Enterococcus* spp., *Aeromonas*, *Cytophaga*, and others; for fungi: *Fusarium* species *Fusarium solani*, *Fusarium oxysporum*, F. *solani*, and *Pseudallescheria boydii Fusarium keratoplasticum*; and viruses manly *Herpesviridae* [29]. In turn, they risk spreading diseases when in contact with other species or populations. Sea turtles are also susceptible to the toxicity and bioaccumulation of environmental pollutants that can affect their health, causing a deterioration in their immune system, and increasing the risk of them developing diseases due to the exposure to pollutants [30, 31].

Therefore, as part of sea turtle protection activities, incorporating a population health assessment program and identifying diseases and abnormalities in wildlife organisms is a priority. However, the lack of knowledge about the physiology of these reptiles makes it difficult to establish criteria to measure the health of their populations and to be able to distinguish between "normal" physiology and the presence of pathologies [30, 32].

Multiple techniques exist to diagnose organism health, including the analysis pollutant concentrations, physical evaluation and disease detection. However, haematological methods are the most widely used as indicators of sea turtle health, since they provide information on immune, cellular and humoral factors, which are necessary for the response to adverse factors, through the blood which can indicate pathological changes in the organisms. Therefore, blood parameters are a non-invasive diagnostic tool that can be used to evaluate and monitor the health status of wildlife [33].

The evaluation of blood parameters, generally accompanied by a detailed clinical history, clinical signs, physical examination and other diagnostic tests, is an important tool for the evaluation of the health of a population through the establishment of blood reference intervals (BRI) and has made it possible to determine and evaluate the functioning of organs and confirm possible diseases in progress. However, in the case of sea turtles, there is difficulty in defining the blood reference values considered normal parameters due to the variability between sea turtles' geographic areas, ecological habitat, populations, sexual maturity, reproductive status, diet and migration. For this reason, therefore, blood parameters must be established for each species and region [31, 33].

Previous studies on the establishment of blood biochemistry reference values in other sea turtle species have found that the values are influenced by ecological implications, the season of the year and the reproductive stage. For example, adult turtles that feed in neritic habitats had low levels of creatinine (Cr), alkaline phosphatase (ALP), Phosphorus (P), Sodium (Na), Magnesium (Mg), unlike adult turtles that feed in ocean habitats [34].

By sex, significant differences have been observed in blood parameters in total protein (PT), albumin (ALB), Calcium (Ca), cholesterol (CHOL), triglycerides (TRIG), cell pack volume (PVC) and total bilirubin (TBIL) being higher in females due to its association with vitellogenesis and folliculogenesis, on the other hand, in males the levels of BUN (blood urea nitrogen) and glucose (GLU) are higher since they are associated with the lack of food during mating. In foraging areas, no significant relationship has been found between the cell packet volume (PCV) with respect to the size and sex of sea turtles, rather PVC is related to feeding and stress on the body. Other parameters such as glucose (GLU) may present higher levels in juveniles than adults, particularly in adult females since they require more energy for the nesting process [35].

Finally, a significant difference has been observed in visibly sick sea turtles compared to healthy individuals, for example in turtles with fibropapillomatosis, the values of total protein (PT), albumin (ALB), cholesterol (CHOL) and triglycerides (TRIG) are lower due to the chronicity and severity of this disease, while in turtles with necrotic or traumatic lesions lower values of total protein (PT), albumin

(ALB), aspartate aminotransferase (AST) and alanine aminotransferase (ALT) have been reported [36].

Some studies on sea turtle blood biochemistry from the northern Gulf of Mexico have included samples from *L. kempii* turtles, which, although the principal objective of these studies was not to establish blood reference intervals (**Table 2**), the results obtained contribute knowledge on the possible differences that the species may present by age or sex, and that factors such as diet do not influence the preprandial and postprandial hematological and plasma parameters in *L. kempii*, unlike other species of sea turtles such as *C. mydas*, in which the biochemical levels of total protein (PT), albumin (ALB), alanine aminotransferase (ALT), alkaline phosphatase (ALP) and postprandial cholesterol (CHOL) increased significantly. The authors mention that this is due to the type of diet of each species, since juvenile *L. kempii* turtles are carnivorous and consume mainly crabs, while juvenile *C. mydas* turtles change their diet from carnivorous to herbivorous when mature [37].

Parameter	**Massachusetts Juveniles Foraging area [12]**	**North Carolina Juveniles Foraging area [38]**	**North Carolina Adults Foraging area [39]**
Sample size (n)	26	10	4
VPC	NA	NA	NA
Total protein (gdL-1)	2.6 ± 0.4	−2.4 – 17.6	2.6 ± 0.9
Albumin (gdL-1)	1.0 ± 0.2	−7.11 – 7.7	0.9 ± 0.3
Globulin (gdL-1)	1.7 ± 0.3	−7.7 – 23.8	1.7 ± 0.6
A/G ratio	0.6 ± 0.1	NA	NA
Total Bilirubin (mgdL-1)	NA	NA	NA
Creatinine (mgdL-1)	0.25 ± 0.11	−33.3 – 50.0	NA
BUN (mgdL-1)	33 ± 22	NA	68.3 ± 20.7
Glucose (mgdL-1)	141 ± 50	−13.1 – 16.9	112.3 ± 48.8
Cholesterol (mgdL-1)	334 ± 141	−2.6 – 10.0	NA
Triglycerides (mgdL-1)	NA	−11.8 – 25.1	NA
Enzymes			
ALKP (UL-1)	285 ± 417	−9.9 – 12.9	NA
ALT (UL-1)	26 ± 50	−100 – 50	NA
AST (UL-1)	610 ± 50	−3.9 – 14.4	108.8 ± 54.9
GGT (UL-1)	3 ± 2	NA	NA
CK (UL-1)	21,979 ± 24,298	−2.4 – 15.8	2,412.3 ± 2,235.4
AMYL (UL-1)	NA		NA
Elements			
Calcium (mgdL-1)	6.6 ± 1.1	−11.5 – 4.5	13.5 ± 9.7
Phosphorus (mgdL-1)	7.4 ± 1.2	−1.1 – 6.9	7.5 ± 4.0

PVC = Packed cell volume, BUN = Blood urea nitrogen, ALKP = Alkaline phosphatase, ALT = Alanine aminotransferase, AST = Aspartate aminotransferase, GGT = Gamma-glutamyl transpeptidase, CK = Creatine phosphokinase, NA = Not analysed.

Table 2.
Reference values of blood biochemistry for different populations of clinically healthy Lepidochelys kempii sea turtles [12, 38, 39].

The impact of gillnets on the sea turtle health, where a significant increase in enzymes such as lactate dehydrogenase (LDH) and creatinine kinase (CK), Phosphorus (P), calcium (K) and glucose (GLU) was observed the longer a turtle was trapped in the net. The authors concluded that the level of stress to which the organism is subjected to alters the blood parameters values. On the other hand, kemps ridley turtles are exposed to pollutants such as metals and organochlorine pesticides from an early age, and even low concentrations of these pollutants are associated with changes in biochemical parameters, which affects the health and behaviour of these organisms [38].

Although the *L. kempii* turtle is considered one of the sea turtle species at the greatest risk of extinction, blood biochemistry reference values have not been established. Therefore, blood biochemistry studies should be prioritised for this species as understanding the health status and disease patterns for these turtles will help secure their future.

4.3 Heavy metal studies

Globally, ocean pollution is a high-impact problem, which is generated from multiple sources and anthropogenic activities, such as fisheries, urban development, mining, production of agricultural and industrial products and oil refineries. These activities significantly increase pollutant levels in marine ecosystems and negatively impact the health of the organisms that inhabit them. Among the primary pollutants found in marine ecosystems are organochlorine compounds, solid waste, plastics (micro and macro), pharmaceuticals, hydrocarbons and heavy metals. Heavy metals are of great importance in ecotoxicology, due to their high toxicity and duration in the environment, since bacteria cannot degrade them over short periods [40–42].

Although some metals are essential for organisms' biochemical processes (Cu, Mn, Mg, Se, Cr, As, Na, K and Mo), high levels of these elements can affect organism health. On the other hand, toxic metals (Hg, Pb and Cd) are elements that are unnecessary in organisms and can alter metabolic pathways and develop diseases and even death [35, 42, 44]. Heavy metals are acquired mainly through diet, and bioaccumulate in specific organs or tissues, such as Cd in the kidney and Pb in bones, by affinity. Through biomagnification processes heavy metals can affect the entire trophic web, with the greatest impact observed in the "top" organisms, such as sea turtles, altering blood biochemistry and other health parameters, since they have the ability to act as endocrine disruptors and produce carcinogenic effects, as well as affecting the fertility of eggs [33, 34, 41].

On the other hand, high concentrations of heavy metals such as Cu, Fe and Pb have been related to the presence of fibropapillomas, a disease which seriously threatens sea turtle health. For this reason, heavy metals such as Cd, Hg and Pb are the pollutants with the greatest impact on these reptiles, even affecting the embryos during vitellogenesis due to the vertical transmission from the female to her eggs; for example, Pb mimics as Ca as its reserves are depleted during nesting seasons and can be transferred to eggs. For these reasons, heavy metals have been identified as a risk to the health of sea turtle populations worldwide.

Although the issue of heavy metals in sea turtles has been widely studied, most studies are based on the analysis of tissues collected from dead individuals. However, this does not increase our understanding on the bioavailability, bioaccumulation and toxicology of heavy metals in sea turtles or their immune response to these pollutants, which is why blood is currently used as analysis tissue, which reflects exposure to short-term pollutants, although it has been observed that there is a relationship between metal levels in the blood with respect to that in other

SCL	Weight	Area	n	Hg	As	Cd	Cu	Pb	Se	Zn	Author
46.9 ± 5.0	16.6 ± 5.0	F	24	0.04 ± 0.04	6.84 ± 1.98*	0.02 ± 0.01*	NA	0.01 ± 0.004*	4.11 ± 1.83*	NA	46
NA	NA	F	106	0.018(0.0005–0.06)	NA	NA	0.52(0.21–1.3)	0.001(0.00–0.03)	NA	7.5(3.28–18.9)	47
46.3 ± 7.1	17.6 ± 7.4	F	18	0.01 ± 0.0092	NA	0.007 ± 0.05	0.470.06	0.02 ± 0.03	NA	3.9 ± 1.47	48
65 ± 3.3	NA	N	18	0.06 ± 0.04	NA	0.01 ± 0.01	0.40 ± 0.09	0.05 ± 0.02	NA	22.70 ± 12.6	48
36.1 ± 7.4	7.4 ± 6.2	F	91	0.01 ± 0.01	NA	0.01 ± 0.005	0.41 ± 0.11	0.03 ± 0.03	NA	6.71 ± 4.46	48

F = Foraging. N = Nesting. NA = Not analysed.
Analysis performed on red blood cells. Mean (min-max) when the standard deviation is not reported.

Table 3.
Heavy metal concentrations reported in different areas (mean ± standard deviation, $\mu g\ g^{-1}$ wet weight) in L. kempii blood [46–48].

tissues such as the liver and kidneys. Another benefit of using blood tissue is that an organism immune response can be observed between the levels of essential (Zn, Cu and Se) and toxic (Hg, Pb and Cd) metals, which suggests that there is a detoxification process by the organism due to the formation of metallothioneins (MT), which are proteins that function as a detoxifying agent.

At present, it is unknown to what degree chemical pollutants are harmful sea turtle health, since the concentration of each metal vary between species and populations, due to the fact that several factors influence the pollutant load, such as habits, diet and the levels of pollutants found in its food, sex, age and physical condition, exposure time, as well as the local habitat characteristics and climatic conditions [40–44]. For this reason, the monitoring of heavy metals in sea turtles must be carried out for each species on a regional and population level.

Sea turtles are sentinel species due to their sensitivity to environmental pollutants which can be analysed in their tissues and reflect the bioaccumulation and behaviour of pollutants such as heavy metals in the environment in which they live. This allows for the identification of potential threats to the environment and health [45].

Regarding the Kemp's ridley turtle (*Lepidochelys kempii*), different authors point out that their population within the Gulf of Mexico is exposed to the heavy metal concentrations present in water and sediments, highlighting that the main sources of contamination in the area are due to large oil spills, such as IXTOC-I in 1979 and Deepwater Horizon in 2010, which increased the concentrations of Pb and other toxic elements, causing potential risks to biota. However, few studies [46] have been carried out on the levels of heavy metals in this sea turtle (**Table 3**), these studies observed that the concentrations of Pb increased with turtle size in a foraging area in Texas, USA, while in another foraging area in Louisiana, USA, Cu and Hg accumulated in higher concentration related to size. On the other hand, this relationship was not observed in the population of the southeast Atlantic. These regional differences were attributed to geographic differences, sources of contamination and levels of metals in the blue crab (*Callinectes sapidus*), the main component of the Kemp's ridley's diet.

These differences observed in Kemps ridley turtles and the interspecific variability of the pollutant load due to the factors described above, show the importance of periodically monitoring the levels of heavy metals in the Gulf of Mexico and in this sea turtle, which would expand knowledge about the levels of these pollutants in the Gulf of Mexico and the impact on the health of this species, contributing to better care and management in *L. kempii* sea turtle conservation programs.

4.4 Molecular studies

The study and analysis of nucleic acids in sea turtles has different approaches which depend on what type of nucleic acid is used to understand what, how and when to conduct molecular analysis. The application of genetics to improve species conservation efforts is an area with great potential, and there is a growing interest in this research.

All organisms or protein entities (such as viruses that contain DNA or RNA inside them) have different nucleic acid sequences (DNA or RNA). Variations are due to factors, such as differences between two individuals, which can be caused due to the time and space where they live, their biology, reproductive success, demographics, places and time of migrations and even natural selection. All the genetic information of each species is stored in its DNA; therefore, if its genome is analysed, it is possible to obtain this information for almost any evolutionary process. In conservation, the acquisition of this knowledge is important for identifying closely related individuals and mitigate inbreeding or exogamy and minimise the loss of genetic variation.

Although taxonomic studies based on morphological characteristics are used to classify individuals or species, molecular analyses are useful in studying sea turtles providing data on their evolution, populations, phylogeny, and how to implement conservation management plans based on molecular results. The study of DNA sequences can provide important information on sea turtle history and data related to their reproductive behaviour and ecology [46]. On the other hand, conservation objectives based on genetics present an opportunity to acquire a greater understanding of a population's status and management for genetic diversity preservation and prevent the persistent risks that affect populations. To better understand sea turtle species, studies should be directed towards those that provide precise data, such is the case of molecular markers.

Thus, molecular markers applied amongst other things to genetic studies, have relevance in countless studies in multiple species, their application in conservation provides information that helps in the understanding of evolutionary history, demography and ecology of endangered species. Information related to species distribution, biology and population dynamics is required to develop efficient and successful conservation strategies. Thanks to the development of molecular analyses and their application in sea turtle conservation, researchers have generated information that has contributed to species recovery [49]. Likewise, the application of molecular techniques can provide a breadth of information on different areas for conservation. However, many techniques present limitations which are subject to their correct use and development, including the correct selection of molecular markers. Therefore, we can consider that genetic or molecular markers are DNA sequences with known locations within a genome, a gene or in a non-coding region. They are generally directly or indirectly associated with the gene's function where they are located or the function of a contiguous gene. Some factors that must be taken into account in the use of molecular markers are the following: variation in the ability to detect differences between individuals, their application costs, ease of use, consistency, multiple capacities (evaluate several loci at the same time) and repetition.

These markers can be small sequences, such as polymorphisms (change or changes of one or more bases within a given sequence, in at least 1% of the population) of a single nucleotide (SNP, single nucleotide polymorphism) or long such as microsatellites [50]. Molecular markers are indispensable tools for determining genetic variation and biodiversity with a high degree of precision and reproducibility.

There are two possibilities within the study of molecular markers: nuclear DNA (nDNA) or in mitochondrial DNA (mDNA) markers. DNA from the mitochondrial genome provides information based on maternal lineage and population dispersion, which is used for analysis in molecular-phylogenetic studies, thanks to its ability to analyse evolution, since it evolves faster than nDNA, resulting in an accumulation of differences between nearby species. The mDNA can estimate gene flow and population history. The information found in the mDNA is conserved, with the absence of introns, short intragenic regions, and few duplications. Markers based on nDNA also called the nuclear genome, provide information on the genetic flow inherited by the male or males, as well as their polygamous reproduction habits with females from different nesting areas, in the same way resolving the paternity question for each nest.

Starting in the 90s, studies on sea turtles using different molecular markers notably increased. These studies have been crucial to increasing our understanding sea turtle ecology. Among the research carried out on sea turtles, one can find species cataloguing and biodiversity inventories [51], identification of illegal sea turtle products, population structure and historical biogeography, phylogenies, female philopatry, male philopatry, multiple paternity, hybridisation, sex ratio [52], epigenetic factors, recombination processes, gene selection and drift that can generate different genealogical histories.

For molecular studies of Kemp's ridley, we should consider different options and marker types. Before going into detail, we need to keep in mind that these studies require a methodical and systematised process, not only for the type of molecular analysis, but during the entire process, from the organisms capture, sampling, DNA extraction, amplification and sequencing and its bioinformatic analysis. Although there are large amounts of information on sea turtle species, molecular studies in the endemic kemps ridley turtle are limited. Therefore, here we will present a general panorama of molecular studies in sea turtles.

Worldwide, biodiversity loss is accelerating due to multiple factors including land exploitation, excessive deforestation, drastic climate change, invasive species and emerging pathogens. Natural resource management focuses on accelerating the inventory of biological diversity, understanding its function and integrating its use in the sustainable development of human society. In this sense, due to the disciplinary crisis in conservation biology and thanks to the evolution of molecular information, a goal was established in which it is necessary to incorporate several technologies to accelerate, ensure and increase the precision of decision-making for conservation. One crucial approach in advancing conservation medicine was centralised database creation, such as the barcode database and DNA records.

In 2003, Paul Hebert and his working group [53] at the University of Guelph, Canada, proposed implementing a coding sequence for cytochrome c oxidase subunit I (COI or CoxI) of the mitochondrial genome as a universal marker for animal species identification. The proposal was based on simulating the universal identifier as a "barcode" used to identify commercial products, their year of manufacture, batch, cost, or simple identification within a warehouse. Thanks to Hebert and his working group's novel work, in May 2004, the Consortium for the Barcode of Life (CBOL) was created, which currently has 130 organisations from 40 countries [54].

Cytochrome c oxidase subunit I (COI, or cox1) is a fragment of the mitochondrial multienzyme complex, which occurs as a transmembrane system related to the mitochondrial matrix. The region used for barcode generation is approximately 648 bp in length. In 2003, the COI gene was designated as standard "barcode" due to its absence of introns, low exposure to recombination, haploid genetic condition, variety of copies that allow easy DNA recovery, and high mutation rate that allows distinction between closely related species, being used as a universal marker for population genetic analysis, identification, phylogeographic studies and cataloguing of all species in all taxa of the animal kingdom. It presents a diversity of nucleotides in certain regions of the gene, a characteristic that allows the discrimination of probable closely related species [55].

The use of this marker with a molecular taxonomic approach was carried out for species-level differentiation studies among 5 sea turtle species. The consumption of sea turtle products and by-products is an activity that often occurs in communities that live close to these species. The application of the "barcode" to identify illegal trafficking of sea turtle meat, eggs, carapace and other by-products helps counteract and combat these illegal activities. It is also useful in identifying stranded animals with a high level of decomposition or when traces of these species are present. Studies have used this mitochondrial fragment to identify samples of hawksbill sea turtle *Eretmochelys imbricata*. In this sense, the use of the mtDNA COI marker is useful for creating phylogenies and population dynamics and contributes directly to conservation actions for sea turtles [50, 56]. Advantages of this marker include the high level of barcode variation between specimens, which helps in the identification of cryptic species and thus its use in species identification, biodiversity and conservation studies. The barcode can also be used to facilitate species identification when this is difficult due to the organism ontogeny or due to only receiving an individual's remains. However, disadvantages exist when using this marker, such as the lack of clarity between the genetic separation values between intra- and interspecific divergence in the selected marker. The discovery of new species cannot be confined solely to a universal barcode, since this marker has errors in the genetic separation values for certain taxonomic groups.

Another commonly used marker is the control region in mDNA, which contains a displacement loop, known as the "D-loop", believed to be the most rapidly evolving region of the mitochondrial genome in most vertebrates. This structure contains most of the regulatory elements for the mitochondrial genome expression and is used to study intraspecific population structures. It has proven useful in the study of sea turtles, with molecular markers derived from the control region used to identify sea turtle natal origins and in the case of the loggerhead turtle (*Caretta caretta*) used to demonstrate their transpacific migration. In 1994, another study designed primers from the control region and amplified a 496 bp fragment for this control region in leatherback turtles [57].

The control region is a ubiquitous characteristic of vertebrate mitochondrial genomes, its name is due to its structure that consists of a triple chain of ~0.8-1 kb and is located between the genes that encode tRNAPro on the light strand (L -strand) and tRNAPhe on the heavy strand (H-strand). This morphology is created by the displacement of the parental H-strand by a DNA of 0.6–0.8 kb complementary to the L-strand. The control region measures between 880 and 1400 bp, and in some species, it can be greatly extended due to its repeated sequences. However, its reduced size and the increased understanding of its replication and transcription mechanisms, its sequence availability in many species, this region represents a good marker to analyse the non-coding regulatory genome to the evolution in the time of the establishment of mammals about 150 million years. In the late 1990s, Dutton et al. described the control region's use, in conjunction with recapture information, to test the philopatry hypothesis. Naro-Maciel et al. conducted a study using the control region as a marker for the DNA barcode, highlighting that although this region is useful in numerous species conservation studies, it does not meet all the criteria for use as a barcode sequence [50, 57].

One advantage of this genetic marker in sea turtle molecular studies is its high degree of variability. It is the most variable region of the mitochondrial and nuclear genome, evolving ten times faster than nDNA and can be used for species identification, even reaching the sub-species level. It degrades slower than nDNA and can be recovered from samples after long storage periods.

5. Conclusion

This chapter showed different stages of sea turtles, one in particular Kemp's ridley. Since that sea turtle has as a unique site to nest, it is important to understand what the real situation about the cycle of life, anthropogenic threats, genetic diversity, and pollution. Not only, for the turtles but for human beings. To see that, previous studies in sea turtles were describe in tissue samples such as carapace, kidney, liver, heart, etc. Another kind of used sample was blood. Blood as test tissue allows information on recent exposure to anthropogenic pollutants, it is inexpensive and easy to acquire. Also, it represents a relationship between the concentrations of heavy metals with other tissues, since the blood transports them through the circulatory system to these. Biochemical analyzes could let us see if some population of this sea turtle is healthier than others o maybe understands if spills, such as the one in 2010 in the Gulf of Mexico, impacts in this species.

On the other hand, there are few molecular studies on these turtles, all of them in secondary nesting areas and with specimens from those places. This chapter would contribute to pointing out molecular studies in mtDNA sequences such as the COI gene, the control region, or in nuclear DNA microsatellite sequences. Which can reveal this species is important to the marine ecosystem conservancy, surely due to its location limited to a single geographical area, which is the Gulf of Mexico. Finally, for many years this species was unknown until brave and clever people decided to act and to protect them. They gave a legacy and showed how marvelous the nesting and hatching are. Now, it is a great opportunity to study with modern tools them and to understand if those sea turtles are healthy and strong to transmit their genes to the next generations of Kemp's ridley. In the end, we showed the state of the art of the smallest, unique, and elusive sea turtle.

Acknowledgements

This research was supported by SIP-20110067, 20161179, 20171851, 20196213, 20200840 and 20211061 projects from the Instituto Politécnico Nacional (IPN). The Apoyos económicos of Comisión de Operación y Fomento de Actividades Académicas (COFAA), Estímulos al Desempeño de los Investigadores (EDI), and Beca de estímulo institucional de formación de investigadores (BEIFI) from the IPN; and the National System of Researchers (SNI); the National Council of Science and Technology (CONACYT) provided fellowships and scholarships for MARL, CPLQ, FYCS, KAZF, VLS, and AAZN. Investigacion, Capacitacion y Soluciones Ambientales y Sociales A.C., Grupo tortuguero de las Californias A. C. We thank the entire community and Juan Martínez Ortíz of Rancho Nuevo, Aldama, Tamaulipas. Special thanks to Martha López Hernández, CONANP, Tamaulipas and Blanca Mónica Zapata Nájera, CONANP., and H. Hugo Acosta Sánchez-United Nations Development Programme-Comisión Nacional de Áreas Naturales Protegidas, Ciudad Victoria, Tamaulipas, Mexico

Conflict of interest

The authors declare no conflict of interest.

Author details

Miguel Angel Reyes-López[1*], Fátima Yedith Camacho-Sánchez[1,2], Catherine E. Hart[3,4], Valeria Leal-Sepúlveda[5], Kevin Alan Zavala-Félix[5], César Paúl Ley-Quiñónez[5], A. Alonso Aguirre[2] and Alan Alfredo Zavala-Norzagaray[5]

1 Conservation Medicine Lab., Centro de Biotecnología Genómica, Instituto Politécnico Nacional, Reynosa, Tam., Mexico

2 Department of Environmental Science and Policy, George Mason University, Fairfax, Virginia, USA

3 Grupo Tortuguero de las Californias, Baja California Sur, Mexico

4 Investigación, Capacitación y Soluciones Ambientales y Sociales AC. Tepic, Nayarit, Mexico

5 Wildlife Lab., Environmental Department, Centro Interdisciplinario de Investigación para el Desarrollo Integral Regional, Unidad Sinaloa, Instituto Politécnico Nacional, Guasave, Sinaloa, Mexico

*Address all correspondence to: mreyesl@ipn.mx; mareyesl@gmail.com

References

[1] Wibbels T., Bevan E. A Historical Perspective of the Biology and Conservation of the Kemp's Ridley Sea Turtle. Gulf of Mexico Science. 2016;33. 129-137. 10.18785/goms.3302.02.

[2] Carr AF. The windward road. Alfred A. Knopf Publishing, New York, 1956; p. 277.

[3] CARR, A. F. 1963. Panspecific reproductive convergence in Lepidochelys kempi. Ergebn. Biol. 26:298-303.

[4] Hildebrand HH. Hallazgo del area de anidacion de la tortuga marina "lora", Lepidochelys kempi (Garman), en la costa occidental del Golfo de Mexico (Rept., Chel.). Ciencia Mexico 1963;22:105-112.

[5] Chavez, H., Contreras-G M., HERNANDEZ-D T. On the coast of Tamaulipas. Int. Turtle. Tortoise Soc. J. 1968;2:20-29.

[6] Heppell SS., Burchfield, PM., Pena LJ. Kemp's ridley recovery, 2007;325-335. In: Biology and conservation of ridley sea turtles. Johns Hopkins University Press, Baltimore, MD.

[7] Caillouet CW. Interruption of the Kemp's ridley population's pre-2010 exponential growth in the Gulf of Mexico and its aftermath: one hypothesis. Mar. Turtle Newsl. 2014;143:1-7.

[8] Crowder L., Heppell S The decline and rise of a sea turtle: how Kemp's ridleys are recovering in the Gulf of Mexico. Solutions. 2011;2:67-73.

[9] Gallaway BJ., Caillouet jr CW., Plotkin PT., Gazey JGC., Raborn SW. 2013. Kemp's ridley stock assessment project. Gulf States Marine Fisheries Commission, Ocean Springs, Mississippi.

[10] CONANP. PROGRAMA DE ACCIÓN PARA LA CONSERVACIÓN DE LA ESPECIE EN RIESGO: Tortuga Lora (*Lepidochelys kempii*). [Internet]. 2021. Available from. https://www.gob.mx/conanp/documentos/programa-de-accion-para-la-conservacion-de-la-especie-en-riesgo-tortuga-lora-lepidochelys-kempii

[11] Bowen BW., Karl SA. Population genetics and phylogeography of sea turtles. Mol Ecol. 2007;16(23):4886-4907. doi: 10.1111/j.1365-294X.2007.03542.x.

[12] Innis C, Nyaoke AC, Williams CR 3rd, Dunnigan B, Merigo C, Woodward DL, Weber ES, Frasca S Jr. Pathologic and parasitologic findings of cold-stunned Kemp's ridley sea turtles (Lepidochelys kempii) stranded on Cape Cod, Massachusetts, 2001-2006. J Wildl Dis. 2009;45(3):594-610. doi: 10.7589/0090-3558-45.3.594.

[13] Wibbels T., Bevan E. *Lepidochelys kempii*. The IUCN red list of threatened species 2019: e.T11533A142050590. Available via: 10.2305/IUCN.UK.2019-2.RLTS.T11533A142050590.en.

[14] Naro-Maciel E., Mihnovets A., Martin M., Durham K., Lii T. Mysteries of an Ancient Mariner: The Endangered Kemp's Ridley Sea Turtle. 2011. Network of Conservation Educators and Practitioners.

[15] Shaver DJ., Hart KM., Fujisaki I., Bucklin D., Iverson AR., Rubio C. . . Ortiz J. Inter-nesting movements and habitat-use of adult female Kemp's ridley turtles in the Gulf of Mexico. PLoS One. 2017;12(3):e0174248. DOI: 10.1371/journal.pone.0174248.

[16] Stanford CB, Iverson JB, Rhodin AGJ, Paul van Dijk P, Mittermeier RA, Kuchling G. . . Walde AD. Turtles and Tortoises

Are in Trouble. Curr Biol. 2020;30(12):R721-R735. doi: 10.1016/j.cub.2020.04.088.

[17] Carreras C. Tortuga lora – *Lepidochelys kempii*. En: Enciclopedia Virtual de los Vertebrados Españoles. Salvador, A., Marco, A. (Eds.). 2013. Museo Nacional de Ciencias Naturales, Madrid. http://www.vertebradosibericos.org/

[18] Caillouet, CW., Gallaway, BJ., Putman NF. "Kemp's Ridley Sea Turtle Saga and Setback: Novel Analyses of Cumulative Hatchlings Released and Time-Lagged Annual Nests in Tamaulipas, Mexico," Chelonian Conservation and Biology, 2016; 15(1):115-131.

[19] Caillouet, CW., Raborn SW., Shaver DJ. Putman NF., Gallaway, BJ., Mansfield KL. Did Declining Carrying Capacity for the Kemp's Ridley Sea Turtle Population Within the Gulf of Mexico Contribute to the Nesting Setback in 2010–2017? Chelonian Conservation and Biology. 2018. DOI: 10.2744/CCB-1283.1

[20] National Marine Fisheries Service & U.S. Fish and Wildlife Service. (2015). Kemp's Ridley Sea Turtle (*Lepidochelys kempii*) 5-Year Review: Summary and Evaluation. repository.library.noaa.gov/view/noaa/17048

[21] Burke RL. Advances in the Biology and Conservation of Turtles. Diversity. 2021; 13(1):17. DOI:10.3390/d13010017

[22] Raposo C., Patrício, A.R., Catry P., Dellinger T., Granadeiro JP. Evidence for trophic differences between live and bycatch oceanic juvenile loggerhead sea turtles. Mar Biol 2019;166, 29 . DOI: 10.1007/s00227-019-3466-8

[23] Hallegraeff GM. Ocean climate change, phytoplankton community responses, and harmful algal blooms: A formidable predictive challenge. Journal of Phycology, 2010;46(2), 220-235. https://doi.org/10.1111/j.1529-8817.2010.00815.x

[24] Pierce RH, Henry MS. Harmful algal toxins of the Florida red tide (Karenia brevis): natural chemical stressors in South Florida coastal ecosystems. Ecotoxicology. 2008;17(7):623-631.

[25] Barrientos RG., Hernández-Mora G., Alegre F., Field T., Flewelling L., McGrath S. . . Stacy, B. A. Saxitoxin Poisoning in Green Turtles (Chelonia mydas) Linked to Scavenging on Mass Mortality of Caribbean Sharpnose Puffer Fish (Canthigaster rostrata-Tetraodontidae). Frontiers in Veterinary Science. 2019. DOI:10.3389/fvets.2019.00466

[26] Ley-Quiñónez C., Zavala-Norzagaray AA., Espinosa-Carreón TL., Peckham H., Marquez-Herrera C., Campos-Villegas L., Aguirre AA. Baseline heavy metals and metalloid values in blood of loggerhead turtles (Caretta caretta) from Baja California Sur, Mexico. Mar Pollut Bull. 2011;62(9):1979-1983. DOI: 10.1016/j.marpolbul.2011.06.022.

[27] Gredzens C and Shaver DJ. Satellite Tracking Can Inform Population-Level Dispersal to Foraging Grounds of Post-nesting Kemp's Ridley Sea Turtles. Front. Mar. Sci. 2020;7:559. DOI: 10.3389/fmars.2020.00559

[28] Shaver DJ, Hart KM, Fujisaki I, Rubio C, Sartain AR, Peña J, Burchfield PM, Gamez DG, Ortiz J. Foraging area fidelity for Kemp's ridleys in the Gulf of Mexico. Ecol Evol. 2013;3(7):2002-2012. DOI: 10.1002/ece3.594.

[29] Mashkour N., Jones K., Kophamel S., Hipolito T., Ahasan S., Walker G... Ariel E. Disease risk analysis in sea turtles: A baseline study to inform conservation efforts. PloS one, 2020;15(10), e0230760.

[30] Aguirre AA., Lutz PL. Marine turtles as sentinels of ecosystem health: Is fibropapillomatosis an indicator? EcoHealth. 2004;1: 275-283.

[31] Ley-Quiñónez CP, Zavala-Norzagaray AA, Réndon-Maldonado JG, Espinosa-Carreón TL, Canizales-Román A, Escobedo-Urías DC, Leal-Acosta ML, Hart CE, Aguirre AA. Selected heavy metals and selenium in the blood of black sea turtle (Chelonia mydas agasiizzi) from Sonora, Mexico. Bull Environ Contam Toxicol. 2013;91(6):645-651. doi: 10.1007/s00128-013-1114-4.

[32] Espinoza-Romo BA, Sainz-Hernández JC, Ley-Quiñónez CP, Hart CE, Leal-Moreno R, Aguirre AA, Zavala-NorzagarayAA.Bloodbiochemistry of olive ridley (*Lepidochelys olivacea*) sea turtles foraging in northern Sinaloa, Mexico. PLoS One. 2018;13(7):e0199825. doi: 10.1371/journal.pone.0199825. PMID: 30044783; PMCID: PMC6059397..

[33] Zavala-Norzagaray AA, Ley-Quiñónez CP, Espinosa-Carreón TL, Canizalez-Román A, Hart CE, Aguirre AA. Trace elements in blood of sea turtles Lepidochelys olivacea in the Gulf of California, Mexico. Bull Environ Contam Toxicol. 2014;93(5):536-541. doi: 10.1007/s00128-014-1320-8.

[34] Ley-Quiñónez CP, Rossi-Lafferriere NA, Espinoza-Carreon TL, Hart CE, Peckham SH, Aguirre AA, Zavala-Norzagaray AA. Associations between trace elements and clinical health parameters in the North Pacific loggerhead sea turtle (Caretta caretta) from Baja California Sur, Mexico. Environ Sci Pollut Res Int. 2017;24(10):9530-9537. DOI: 10.1007/s11356-017-8556-x.

[35] Camacho M, Orós J, Boada LD, Zaccaroni A, Silvi M, Formigaro C, López P, Zumbado M, Luzardo OP. Potential adverse effects of inorganic pollutants on clinical parameters of loggerhead sea turtles (Caretta caretta): results from a nesting colony from Cape Verde, West Africa. Mar Environ Res. 2013;92:15-22. DOI: 10.1016/j.marenvres.2013.08.002.

[36] Prieto-Torres DA., Hernández JL., Bravo Henríquez AR., Alvarado MC., Dávila MJ. "Blood Biochemistry of the Breeding Population of Green Turtles (Chelonia mydas) in the Aves Island Wildlife Refuge, Venezuela," South American Journal of Herpetology, 2013;8(3), 147-154. DOI:10.2994/SAJH-D-13-00010.1.

[37] Brenes-Chaves L., Berrocal A., Meneses A., Sánchez C., Vásquez C. Study on the etiology of fibropapillomatosis of olive ridley sea turtles (*Lepidochelys olivacea*) nesting in the National Wildlife Refuge at Ostional, Guanacaste, Costa Rica. Revista Ciencias Marinas y Costeras. 2013;5. 119. 10.15359/revmar.5.8.

[38] Anderson ET., Minter LJ., Clarke EO 3rd., Mroch RM 3rd., Beasley JF., Harms CA. The Effects of Feeding on Hematological and Plasma Biochemical Profiles in Green (Chelonia mydas) and Kemp's Ridley (Lepidochelys kempii) Sea Turtles. Vet Med Int. 2011;2011:890829. DOI: 10.4061/2011/890829.

[39] Snoddy, JE., Landon M., Blanvillain G., Southwood A. Blood biochemistry of sea turtles captured in gillnets in the Lower Cape Fear River, North Carolina, USA. Journal of Wildlife Management. 2009;73 (8): 1394-1401.

[40] Cortes-Gomez AA., Romero D., Santos J., Rivera-Hernandez JR., Girondot M. Inorganic elements in live vs dead nesting olive ridley marine turtles in the Mexican Pacific: Introducing a new statistical methodology in ecotoxicology. Sci Total Environ, 2020;143249. DOI: 10.1016/j.scitotenv.2020.143249

[41] Aguirre AA., Gardner SC., Marsh JC., Delgado SG., Colin JL., Wallace JN. Hazards Associated with the Consumption of Sea Turtle Meat and Eggs: A Review for Health Care Workers and the General Public. EcoHealth, 2006;3, 141-153. DOI: 10.1007/s10393-006-0032-x

[42] Caron A. G. M., Thomas C. R., Berry K. L. E., Motti C. A., Ariel E., Brodie, J. E. Ingestion of microplastic debris by green sea turtles (Chelonia mydas) in the Great Barrier Reef: Validation of a sequential extraction protocol. Mar Pollut Bull, 2018;127, 743-751. DOI:10.1016/j.marpolbul.2017.12.062

[43] Rai P. K., Lee S. S., Zhang M., Tsang Y. F., Kim K. H. Heavy metals in food crops: Health risks, fate, mechanisms, and management. Environ Int, 2019;125, 365-385. DOI:10.1016/j.envint.2019.01.067

[44] Mühlenbruch M, Grossart HP, Eigemann F, Voss M. Mini-review: Phytoplankton-derived polysaccharides in the marine environment and their interactions with heterotrophic bacteria. Environ Microbiol. 2018;20(8):2671-2685. DOI: 10.1111/1462-2920.14302.

[45] Perrault J. R., Stacy N. I., Lehner A. F., Mott, C. R. Hirsch S., Gorham J. C., . . . Walsh, C. J. Potential effects of brevetoxins and toxic elements on various health variables in Kemp's ridley (*Lepidochelys kempii*) and green (*Chelonia mydas*) sea turtles after a red tide bloom event. Sci Total Environ, 2017;605-606, 967-979. DOI:10.1016/j.scitotenv.2017.06.149

[46] Orvik, LM. Trace metal concentration in blood of the Kemp's ridley sea turtle (*Lepidochelys kempii*). Master's thesis, Texas A&M University. 1997. Available electronically from http://hdl.handle .net/1969.1/ETD-TAMU-1997-THESIS-O78.

[47] Wang, HC. Trace metal uptake and accumulation pathways in Kemp's ridley sea turtles (*Lepidochelys kempii*). Doctoral dissertation, Texas A&M University. Texas A&M University. 2006. Available electronically from http://hdl.handle.net/1969.1/2413.

[48] Naro-Maciel E., Reid, B., Alter S., Amato G., Bjorndal K., Bolten A. . . Pineda-Catalan O. From refugia to rookeries: Phylogeography of Atlantic green turtles. Journal of Experimental Marine Biology and Ecology. 2014. 306-316. 10.1016/j.jembe.2014.08.020.

[49] Duchene S, Frey A, Alfaro-Núñez A, Dutton PH, Thomas P Gilbert M, Morin PA. Marine turtle mitogenome phylogenetics and evolution. Mol Phylogenet Evol. 2012;65(1):241-250. DOI: 10.1016/j.ympev.2012.06.010.

[50] Al-Samarai F., Al-Kazaz, A. Molecular Markers: an Introduction and Applications. European Journal of Molecular Biotechnology. 2015;9. 10.13187/ejmb.2015.9.118CrossRef

[51] Naro-Maciel E, Le M, FitzSimmons NN, Amato G. Evolutionary relationships of marine turtles: A molecular phylogeny based on nuclear and mitochondrial genes. Mol Phylogenet Evol. 2008 Nov;49(2):659-662. DOI: 10.1016/j.ympev.2008.08.004.

[52] Lasala JA, Hughes CR, Wyneken J. Breeding sex ratio and population size of loggerhead turtles from Southwestern Florida. PLoS One. 2018;13(1):e0191615. DOI: 10.1371/journal.pone.0191615.

[53] Hebert PD, Cywinska A, Ball SL, deWaard JR. Biological identifications through DNA barcodes. Proc Biol Sci. 2003;270(1512):313-321. DOI: 10.1098/rspb.2002.2218.

[54] Hebert PD, Hollingsworth PM, Hajibabaei M. From writing to reading the encyclopedia of life. Philos Trans R Soc Lond B Biol Sci.

2016;371(1702):20150321. DOI: 10.1098/rstb.2015.0321.

[55] Hebert PDN, Braukmann TWA, Prosser SWJ, Ratnasingham S, deWaard JR, Ivanova NV, Janzen DH, Hallwachs W, Naik S, Sones JE, Zakharov EV. A Sequel to Sanger: amplicon sequencing that scales. BMC Genomics. 2018;19(1):219. DOI: 10.1186/s12864-018-4611-3.

[56] Naro-Maciel E, Reid B, Fitzsimmons NN, LE M, Desalle R, Amato G. DNA barcodes for globally threatened marine turtles: a registry approach to documenting biodiversity. Mol Ecol Resour. 2010;10(2):252-263. DOI: 10.1111/j.1755-0998.2009.02747.x. Epub 2009 Aug 20. PMID: 21565020.

[57] Lahanas PN, Miyamoto MM, Bjorndal KA, Bolten AB. Molecular evolution and population genetics of Greater Caribbean green turtles (Chelonia mydas) as inferred from mitochondrial DNA control region sequences. Genetica. 1994;94(1):57-66. DOI: 10.1007/BF01429220. PMID: 7729697.

Section 2

Applied and Human Ecology

Chapter 7

Fish and Fisheries of the Eastern Coast of Mexico, with Emphasis on Coral Reef Species

Carlos González-Gándara and Ernesto A. Chávez

Abstract

The state on knowledge of fish communities associated with coral reefs of the southern Gulf of Mexico (Veracruz, Campeche bank), and eastern Yucatan on the Caribbean is reviewed, in addition to a description of the main fisheries of the area. The review includes coral reef fish of Veracruz, the Campeche Bank, and reefs running along the Caribbean coast up to the border with Belize. Data recorded suggest that the heterogeneity of different levels (region, reef and reef zone) may be responsible for a larger number of niches available, promoting higher specific diversity that is more evident in the Caribbean reefs. The environmental conditions create patterns of differential abundance among the three zones. The main regional fisheries include more than 60 species and the current yield suggests a 30% reduction compared to catch volumes recorded a few years ago. The changes in coral coverage and the fishing pressure over coral reefs have exerted effects on species of fishing importance.

Keywords: coral reefs, fish community, southern Gulf of Mexico, Mexican Caribbean

1. Introduction

The fish fauna of the south Gulf of Mexico, the Campeche Bank and the Caribbean coast of Yucatan Peninsula comprise the northernmost extreme of the tropical sea, included in the region known as the Caribbean province in the tropical belt of Eastern America. This region has a high species diversity combing a marine shelf covered by terrigenous sediments and coral reefs, allowing the possibility to maintain a diverse fish community, supporting the statement that between 66 and 89% of marine fish species are dwellers of coral reefs and reef-associated habitats [1].

Ichthyofaunal research in the reefs of the Gulf of Mexico and the Mexican Caribbean is just over 50 years old. The first formal documents allude to the components of Alacranes reef, Yucatan [2], Blanquilla reef, Veracruz [3] and Cozumel, Quintana Roo [4]. However, the knowledge of the ichthyological components of both the southern Gulf of Mexico and the Mexican Caribbean is incomplete, because sampling efforts have been isolated and dispersed over time [5]. Among the works that synthesize the regional ichthyofauna stand out Díaz-Ruiz et al., Schmitter-Soto, and Schmitter-Soto et al. [5–7] for the Mexican Caribbean, as well as Chávez and Beaver [8] for reef systems of the southern Gulf of Mexico.

The publications that have addressed the structure of fish communities (richness, distribution, relationship with environmental factors, etc.) are numerous, some of them correspond to particular systems, such as the Veracruz Reef System [9], Cozumel [10]; or Chinchorro Bank [11]. Others include a general analysis of the reef fish communities of the Gulf of Mexico [12] and the Caribbean [13–15]. Fishing contributes to degradation of coral reefs and the fisheries associated to these ecosystems, includes a smaller component of coral reef dwelling species [16, 17]. This chapter summarizes the information published so far, comparing the ichthyological components and their relative abundance. In addition, an assessment of the most important fisheries associated with the reef structures of the southern Gulf of Mexico and the Mexican Caribbean is made.

2. The habitats

The eastern Gulf of Mexico and the Mexican coast of the Caribbean Sea are a heterogeneous region where three subzones can be defined by the characteristics of the habitat, the western Gulf, the Yucatan platform, also known as Campeche Bank, and the Caribbean coast of the Yucatan peninsula [18]. Despite these regions hold some differential characteristics, the fish fauna has many similarities, and despite the Caribbean coast holds essentially a coral reef habitat, there are many common fish species along these areas. It is considered that more than 100 fish species are closely associated to coral reefs and hard bottoms [8], so preservation of integrity of these communities is important to the economic and ecological health of this region.

The south Gulf of Mexico, from the border with the USA, is considered of tropical nature, where two main habitats are well defined, one which may be subdivided into two main ecosystem types, the first one is shrimp grounds, strongly associated to coastal lagoons and mangroves; the other one is the coral reef ecosystems divided in two subregions (Veracruz and Campeche reefs). The coast of Veracruz contains more than 100 coral reef structures [19] divided in three systems: the Lobos Tuxpan Reef System (LTRS), the Veracruz Reef System (VRS) and the Tuxtlas Reef System (TRS) [20]. Most reefs in Veracruz develop under stress conditions (turbidity and thermal stress) due to their proximity to the coast [21] and the drainage of the ten hydrological basins that flow towards the Veracruz coast [22]. According to their geomorphology, the Veracruz's reefs are: platform, fringing and submerged banks. Zonation of the platform-type reefs (crest, lagoon, windward and leeward slopes) is defined by the benthic components [21, 23–26] resulting from wave exposure, ocean currents, suspended sediments, and turbidity [27, 28]. Coral coverage of these reefs is 15 to 25% [21, 26]. Veracruz's submerged banks have an irregular to oval or semicircular shape, with a depth ranging from 1 to 40 m [19]. Although there are few studies, the coral coverage of these reefs is less than 10% [29, 30], excepting Blake reef, whose coral coverage is higher than 15% [31]. Fringing reefs are small formations (<1.0 km long) in central Veracruz, whose coral cover is less than 10% [32]. In the south, there is a strip of approximately 3.5 km [33] (**Figure 1**, **Table 1**).

The coral formations of the Campeche Bank (CBRS) are geomorphologically divided in two: platform-type or emergent (Arcas) and submerged banks (Banco Pera), among them, Alacranes reef (22 km long), is on the north of Yucatan peninsula, at 150 km from the coast [34] (**Figure 1**, **Table 1**). Jordán-Dahlgren [35] refers to up to 46 structures in this region. Platform-type reefs present a similar zoning to that observed in Veracruz reefs [24, 25]. Their depth ranging from 1 to 44 m [35, 36].

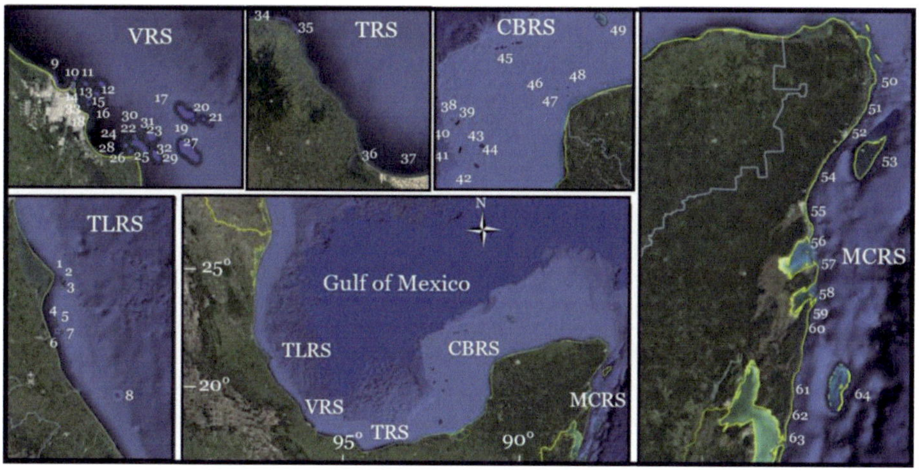

Figure 1.
Coral reefs of the south Gulf of Mexico and Mexican Caribbean. 1=Blanquilla, 2=Medio, 3=Lobos, 4= Tanhuijo, 5=Enmedio, 6=Pantepec, 7=Tuxpan, 8=Blake, 9=Punta Gorda, 10=Galleguilla, 11=Blanquilla, 12=Anegada de Adentro, 13=Verde, 14=Hornos, 15=Pájaros, 16=Sacrificios, 17=Anegada de Afuera, 18= Mocambo, 19=Topatillo, 20=Santiaguillo, 21=Anegadilla, 22=Polo, 23=Enmedio, 24=Los Bajitos, 25=Chopas, 26=Blanca, 27=Cabezo, 28=Giote, 29=Rizo, 30=La Palma, 31=Los Sargazos, 32=Periférico, 33=Terranova, 34=La Perla, 35=Zapotitlán, 36=Palo Seco, 37=Tripie, 38=Triángulos Oeste, 39=Triángulos Este, 40=Obispo Norte, 41=Obispo Sur, 42=Arcas, 43=Banco Pera, 44=Banco Nuevo, 45=Arenas, 46=Serpiente, 47=Madagascar, 48=Sisal, 49=Alacranes, 50=Punta Nizuc, 51=Puerto Morelos, 52=Punta Maroma, 53=Cozumel, 54= Akumal, 55=Boca Paila, 56=Yuyum, 57=Punta Allen, 58=Punta Herrero, 59=Tampalam, 60=El Placer, 61=Mahahual, 62=Xahuayxol, 63=Xcalak, 64=Chinchorro.

The highest part of the submerged banks is between 3 and 15 m below sea level and its maximum depth is 18 m [37, 38] and the coral cover of these reefs ranges from 5 to 25% [25]. The coral reefs located on the west shelf of the Campeche Bank, play the role of a source of biodiversity of an important coral and fish community which supply with larval stages to the coral reef ecosystem of the Veracruz shelf [39, 40].

Finally, the Caribbean coast of Yucatan Peninsula almost lacks of a continental shelf, and the coast is profusely occupied by fringing reefs, from Isla Contoy to the border with Belize. This region also includes insular structures (Cozumel and Banco Chinchorro) and submerged banks (Arrowsmith) [35, 41]. The fringing formations grow parallel to the coast or are separated by a well-defined lagoon. These systems are divided into three major groups: the southern group is made up of well-developed reefs that form channel systems and massif crest reefs whose geographic limit is the 19°05′ coordinate, the central-northern group, which goes from the previous geographic limit to Contoy Island, with well-developed structures in the front, is dominated by gorgonians and algae, and the deep reefs that are located in southwest Cozumel with coral development between 10 and 50 m depth [24] (**Figure 1**, **Table 1**). The coral cover of most reefs in this region ranges from 15 to 20% [41].

Zonation of coral reefs of the Mexican Caribbean is related to environmental factors, like wave impact and light penetration [42] and include: a shallow lagoon covered with sand, seagrasses, macroalgae and scarce coral colonies; a shallow back reef covered with coral fragments and massive coral colonies; a shallow reef crest exposed to surf that is covered by branching corals (*Acropora palmata*), fire coral (*Millepora* spp) and coralline algae; a reef front with a gentle slope facing the prevailing winds, with a depth of 5 to 25 m. This zone is covered with branching and massive corals, sand and debris, and the reef slope is covered with massive and foliose corals as well as sponges and gorgonians. This zone is an extension of the reef front with a high slope at a depth of 12 to 40 m [13, 14] or more.

Reef	Geographic location	Average surface (ha)	Depth (m)
1. Blanquilla	21°32′33″N; 97°16′49″W	200.76	0.5–20
2. Medio	21°30′44″N; 97°15′09″W	118.78	0.5–20
3. Lobos	21°28′19″N; 97°13′27″W	398.26	0.5–30
4. Tanhuijo	21°08′17″N; 97°16′18″W	155.45	0.5–20
5. Enmedio	21°04′56″N; 97°15′20″W	237.05	0.5–20
6. Pantepec	21°02′40″N; 97°14′27″W	100.99	16–27
7. Tuxpan	21°01′44″N; 97°11′43″W	144.99	0.5–30
8. Blake	20°45′47″N; 96°59′24″W	124.49	9–34
9. Punta Gorda	19°15′05″N; 96°10′45″W	39.03	0–4
10. Galleguilla	19°13′49″N; 96°07′22″W	34.99	0–18
11. Blanquilla	19°13′35″N; 96°05′51″W	42.49	0–24
12. Anegada de Adentro	19°13′33″N; 96°03′19″W	75.85	0–36
13. Verde	19°12′11″N; 96°04′03″W	67.11	0–27
14. Hornos	19°11′28″N; 96°07′13″W	12.02	0–3
15. Pájaros	19°11′18″N; 96°05′21″W	113.08	0–18
16. Sacrificios	19°10′35″N; 96°05′31″W	45.18	0–14
17. Anegada de Afuera	19°09′23″N; 95°51′23″W	471.6	0–45
18. Mocambo	19°09′00″N; 96°05′25″W	43.73	0–7
19. Topatillo	19°08′30″N, 95°50′08″W	13.72	0–45
20.Santiaguillo	19°08′29″N; 95°48′30″W	17.5	0–45
21. Anegadilla	19°08′09″N; 95°47′43″W	20.62	0–45
22. Polo	19°06′29″N; 95°58′37″W	21.97	0–24
23. Enmedio	19°06′21″N; 95°56′20″W	258.62	0–24
24. Los Bajitos	19°06′06″N; 95°58′28″W	42.79	
25. Chopas	19°05′22″N; 95°58′07″W	473.63	0–24
26. Blanca	19°05′10″N; 95°59′56″W	41.33	0–18
27. Cabezo	19°04′31″N; 95°51′00″W	1037.59	0–24
28. Giote	19°04′08″N; 95°59′55″W	3.46	0–2
29. Rizo	19°03′50″N; 95°55′41″W	184.41	0–18
30. La Palma	19°07′15″N; 95°57′58″W	197.14	4–26
31. Los Sargazos	19°06′22″N; 95°56′47″W	19.22	2–22
32. Periférico	19°04′57″N; 95°56′03″W	4.59	2–16
33. Terranova	19°10′59″N; 96°05′42″W	6.53	1–15
34. La Perla	18°32′35″N; 94°49′34″W	ND	1–12
35. Zapotitlán	18°27′42″N; 94°45′44″W	ND	1–12
36. Palo Seco	18°10′33″N; 94°31′32″W	ND	11–22
37. Tripie	18°10′33″N; 94°22′03″W	ND	7–15
38. Triángulos Oeste	20°57′58″N; 92°17′56″W	ND	0–18
39. Triángulos Este	20°54′31″N; 92°12′55″W	ND	0–18
40. Obispo Norte	20°28′41″N; 92°12′21″W	ND	5–18
41. Obispo Sur	20°25′28″N; 92°13′25″W	ND	5–18

Reef	Geographic location	Average surface (ha)	Depth (m)
42. Arcas	20°11′47”N; 91°57′58”W	ND	0–18
43. Banco Pera	20°42′55”N; 91°55′05”W	ND	16–44
44. Banco Nuevo	20°30′00”N; 91°50′44”W	ND	0–30
45. Arenas	22°06′42”N; 91°23′27”W	ND	0–30
46. Serpiente	20°57′58”N; 92°17′56”W	0.21	7–18
47. Madagascar	20°57′58”N; 92°17′56”W	0.21	4–13
48. Sisal	20°57′58”N; 92°17′56”W	0.67	3–10
49. Alacranes	22°28′43”N; 89°42′05”W	333,768	0.5–50
50. Punta Nizuc	21°19′10”N; 86°46′30”W	ND	1- > 30
51. Puerto Morelos	20°51′16”N; 86°51′40”W	9066	0.5–25
52. Punta Maroma	20°43′42”N; 86°57′37”W	ND	1- > 30
53. Cozumel	20°25′16”N; 86°55′52”W	11,987	1.5- < 40
54. Akumal	22°23′24”N; 87°18′30”W	ND	1.5–65
55. Boca Paila	20°01′11”N; 87°28′06”W	ND	1- > 30
56. Yuyum	19°54′49”N; 87°25′42”W	ND	1–45
57. Punta Allen	19°46′20”N; 87°26′52”W	ND	1- > 30
58. Punta Herrero	19°19′42”N; 87°26′43”W	ND	1- > 30
59. Tampalam	19°06′54”N; 87°32′11”W	ND	1- > 30
60. El Placer	18°53′23”N; 87°38′06”W	ND	1- > 30
61. Mahahual	18°42′49”N; 87°42′07”W	ND	1- > 30
62. Xahuayxol	18°21′15”N; 87°47′27”W	ND	1- > 30
63. Xcalak	18°15′35”N; 87°49′25”W	17,949	1–70
64. Chinchorro	18°34′24”N; 87°21′06”W	144,360	1.5–60

Table 1.
Geographic location and characteristics of the coral reefs from the South Gulf of Mexico and Mexican Caribbean. Source of data [19, 20, 25, 37, 38, 43–48].

3. Fish communities

As a consequence of habitat morphology, the study of coral reef fish is based upon visual censuses. This way, it makes not easy to compare abundance data of coral reefs with fish faunas caught with trawl and other kind of nets on soft grounds. However, quantitative data provide records of relative abundance and therefore allows to get a picture of fish communities in each habitat.

The current stream on the Mexican Caribbean is one of the factors determining the physical and chemical properties of the marine waters of the Gulf of Mexico [39] and is a decisive factor in the ichthyologic composition [12]. The current system, the rainfall regime and the continental drainage modify the environmental conditions at the regional level [49], leading to the presence of endemic species and influencing the abundance of some others. The fish community associated to the reefs of the southern Gulf of Mexico and the Mexican Caribbean is made up of 776 species belonging to 115 families. The highest species richness corresponds to the Caribbean (579 species) [4, 6, 10, 11, 50], followed by the reefs of Veracruz (509 species) [3, 9, 32, 33, 51–54] and the Campeche Bank (445 species) [2, 12, 38, 55, 56]. The differences are related in part to the sampling effort carried out and the

Family	Veracruz Coast	Campeche Bank	Mexican Caribbean	Total
Serranidae	45	39	45	63
Gobiidae	27	30	22	48
Carangidae	23	20	24	25
Labrisomidae	10	16	22	24
Haemulidae	20	15	16	21
Sciaenidae	19	8	7	19
Labridae	16	16	16	19
Pomacentridae	15	15	14	15
Scaridae	14	14	14	14

Table 2.
Species richness of the main fish fauna associated to the reefs of the South Gulf of Mexico and the Mexican Caribbean.

distance to the center of origin [57], since most of them are of Caribbean origin. On the other hand, regional conditions (turbidity, nutrient concentration, etc.) are determining factors in the fish components [27]. For example, in Veracruz, species that normally inhabit estuarine areas (e.g. *Bairdiella veraecrucis*) have been recorded in the proximity of the reefs located near the coast [54]. Published data for the reefs of the southern Gulf of Mexico and the Mexican Caribbean [2, 4, 6–13, 32, 33, 38, 50–56] reveal that the families with the highest species richness are: Serranidae, Gobiidae, Carangidae, and Labrisomidae (**Table 2**).

Regional conditions of salinity, temperature, concentration of nutrients, among others, as well as the heterogeneity of the reef environment, create patterns of differential abundance among the three regions. Thus, the proportions estimated from published [11–13, 32, 58, 59] and unpublished data show that, Blue tang (*Acanthurus coeruleus*), Blue chromis (*Chromis cyanea*), Yellowhead wrasse (*Halichoeres garnoti*) and Bluehead (*Thalassoma bifasciatum*) are higher in the Caribbean, while Glass goby/Masked goby (*Coryphopterus hyalinus-personatus*), Mardi Gras wrasse (*Halichoeres burekae*), Doctorfish (*Acanthurus chirurgus*) and Brown chromis (*Chromis multilineata*) show a higher relative abundance in the reefs of the Gulf of Mexico (**Table 3**).

	Veracruz Coast	Campeche Bank	Mexican Caribbean
Abudefduf saxatilis	2–5%	<1%	2–5%
Acanthurus chirurgus	<1%	<1%	1–2%
Acanthurus coeruleus	<1%	<1%	2–5%
Acanthurus tractus	<1%	<1%	2–5%
Bodianus rufus	2–5%	<1%	
Chromis cyanea	<1%	<1%	10–20%
Chromis multilineata	10–20%	1–2%	1–2%
Chromis scotti	2–5%	2–5%	
Clepticus parrae	<1%	<1%	5–10%
Coryphopterus hyalinus-personatus	20–30%	20–30%	<5%
Elacatinus jarocho	2–5%		
Gramma loreto		1–2%	1–2%
Haemulon aurolineatum	5–10%	1–2%	<1%

	Veracruz Coast	Campeche Bank	Mexican Caribbean
Haemulon flavolineatum	<1%	2–5%	2–5%
Haemulon sciurus			2–5%
Halichoeres bivittatus	2–5%	5–10%	1–2%
Halichoeres burekae	>30%	2–5%	<1%
Halichoeres garnoti	<1%	1–2%	2–5%
Lujanus apodus			1–2%
Microspathodon chrysurus	<1%	<1%	1–2%
Ocyurus chrysurus	<1%	1–2%	
Scarus iseri	2–5%	10–20%	2–5%
Scarus taeniopterus			1–2%
Sparisoma aurofrenatum	<1%	2–5%	2–5%
Sparisoma viride	<1%	1–2%	2–5%
Stegastes adustus	2–5%	1–2%	1–2%
Stegastes partitus	1–2%	5–10%	5–10%
Stegastes planifrons	1–2%	2–5%	1–2%
Stegastes xanthurus	2–5%	1–2%	<1%
Thalassoma bifasciatum	2–5%	2–5%	10–20%

Table 3.
Relative abundance of the most abundant fish species of coral reefs of the South Gulf of Mexico and the Mexican Caribbean.

Environmental heterogeneity has been referred to as a cause of ichthyofaunal diversity [60, 61] and this heterogeneity can be analyzed at various scales: region, reef and reef zone. At the region level, oceanographic processes (currents, gyres, etc.) determine the flow of nutrients and larvae to the reefs [62]. This, together with the availability of physical spaces for recruitment and settlement of larvae as well as the availability of food, participate in the composition and abundance of fish. The conditions of higher environmental stability in the Mexican Caribbean seem to be related to its greater richness. However, the flow of fresh water and sediments on the reefs of Veracruz promotes a mixing of estuarine and marine fishes.

4. Heterogeneity and fish richness

At the coral reef scale, the geomorphology, the dimensions of the reef, the depth and its distance to the coast, as well as local current patterns, temperature regime, and salinity, are some of the factors defining the structure of fish communities [8, 28, 63]. At this scale, heterogeneity is related to the dimensions of each reef [25] which diversifies the niches and enhancing specific diversity, for example, in Alacranes reef (Gulf of Mexico) and Chinchorro Bank (Caribbean). The fish richness of the fringing reefs of the Caribbean, is higher in the southern reefs (Mahahual, Yuyum, and Boca Paila) compared to the northern reefs (Punta Nizuc, Puerto Morelos, and Punta Maroma) [15], perhaps because deep fore reefs tend to be smaller and less developed in the north [41], while in the center and southern, the massive coral species appear to determine the structure and diversity of the coral assemblages, influencing fish aggregations [15]. The **Figure 2** show the fish richness in some reefs of the south Gulf of Mexico and the Mexican Caribbean [2, 4, 6–13, 32, 33, 38, 50–59].

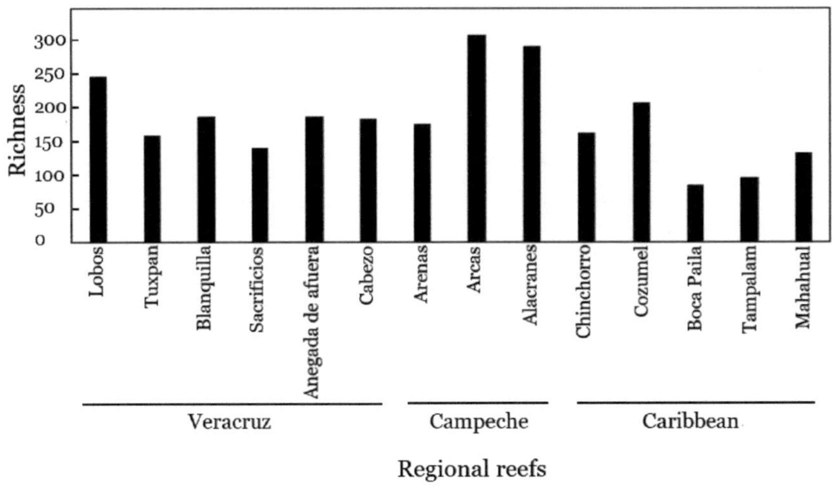

Figure 2.
Ichthyologic species richness recorded at some reefs of the southern Gulf of Mexico and the Mexican Caribbean.

The ichthyofaunal components of near-shore reef structures may be influenced by the fauna that inhabit freshwater or estuarine systems. Reefs of central Veracruz that are very close to the mouths of the Jamapa and La Antigua rivers have typically estuarine species, such as: *B. veraecrucis* [54]. This connection is also observed in Tamiahua Lagoon, Veracruz and Términos Lagoon, Campeche, where typical reef fish species like *Aluterus schoepfii* and *Stephanolepis hispidus* have been detected respectively [64, 65]. This proximity provides nutrients to reef areas and limits light penetration due to suspended sediments, which modifies the abundance of some species, such as *T. bifasciatum* [66].

In the Caribbean, reefs connected to rivers running underground, have components cataloged as freshwater, such as *Mayaheros urophthalmus* [67] or estuarine species, such as *Mugil* sp. [68]. On fringing reefs, some fishes (e.g. Lutjanidae, and Haemulidae) move from the reef to the mangrove in search of food or for breeding [69, 70], explaining their high abundance values recorded in the Caribbean reefs especially in shallow areas [13, 68].

At reef zone level, exposure to dominant winds, water transparency, depth and benthic coverage define fish richness and distribution [28, 66]. In the platform-type reefs of Veracruz and Campeche, the greater coral coverage in the leeward slope favors the richness and abundance of fish, particularly those that live closely associated with coral structures. In this reef zone are common *Halichoeres burekae*, *Coryphopterus hyalinus*, *Haemulon aurolineatum*, *Stegastes planifrons*, *Chromis multilineata*, *Chromis scotti*, *Scarus iseri*, and *Sparisoma aurofrenatum*. On the windward slope, structural complexity as well as currents concentrate plankton are determinant in the fish community structure. In the deep areas, are frecuent *H. burekae*, *C. multilineata*, *C. hyalinus,* and *Abudefduf saxatilis*. In the shallow areas of leeward, *Stegastes adustus*, *Ophioblennius macclurei*, *T. bifasciatum*, and *Microspathodon chrysurus* are the most abundant. In the crest, the fish species adapted to swell like *S. adustus, A. saxatilis*, *O. macclurei*, *T. bifasciatum,* and *M. chrysurus* dominate. Finally, in the lagoon reef, species like *Halichoeres bivittatus*, *S. adustus*, *Stegastes leucostictus* and *Sparisoma radians* are common. **Figure 3** shows the fish richness by reef zone, estimated from unpublished data on reefs from northern Veracruz, the average richness is higher on the slopes.

In the Mexican Caribbean, the average ichthyologic species richness is higher in the areas exposed to the waves (slope and terraces) in comparison to the lagoon

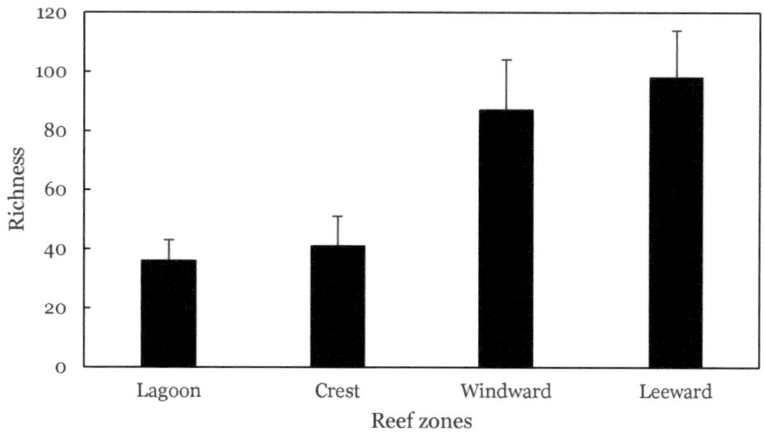

Figure 3.
Mean fish species richness by reef zone at coral reefs of northern Veracruz, Mexico.

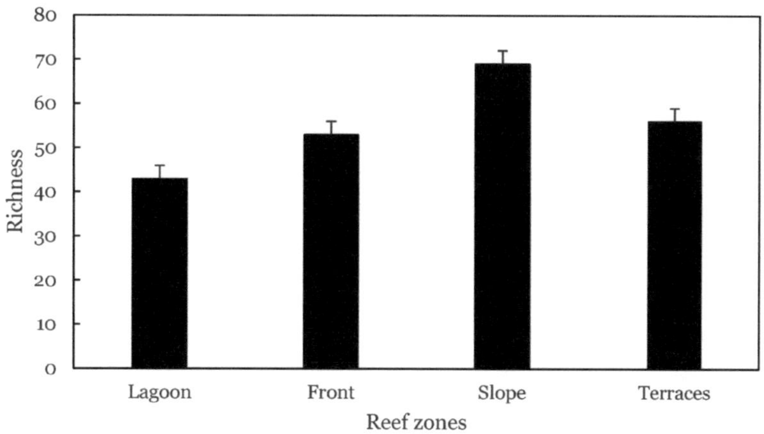

Figure 4.
Mean ichthyologic richness by reef zone in Mexican Caribbean reefs. Data sources [13, 15].

and the reef front (**Figure 4**). Distribution of fish species is fitted to the availability of microhabitats as well as to the environmental preferences of each species. In the lagoons, the juvenile stages of *Haemulon sciurus*, *Haemulon flavolineatum*, *Lutjanus apodus*, *Ocyurus chrysurus*, and *Acanthurus chirurgus*, among others, are frequent. In addition, species linked to sea grasses and sand (*S. radians*, *Eucinostomus lefroyi*, *Gerres cinereus*, and *H. bivittatus*) are common too. On the reef crest, where the substrate is covered by branched corals, calcareous algae and turf algae, are common the species adapted to intense wave action as *S. adustus*, *M. chrysurus,* and *A. saxatilis*. In addition, schools of *Kyphosus vaigiensis*, *Acanthurus tractus*, *A. chirurgus*, and *S. iseri* are transient through this area. On the reef slope, schools of fish that take advantage of the concentration of zooplankton generated by the currents can be observed. Among these species, stand out for their abundance *C. scotti*, *Chromis cyanea*, and *Clepticus parrae*. In the interspaces produced by the coral development, fish species that use the spaces between corals to protect themselves from predators (*Haemulon melanorum*, *Haemulon macrostomum*) and species that use these spaces to go unnoticed by the prey (*Epinephelus morio*) are often seen. Also common are species that transit the water column in search of food (*Caranx ruber*, and *Lutjanus cyanopterus*). In the terraces characterized by the presence of gorgonids, *Cephalopholis fulva*, *Epinephelus guttatus*, and *Stegastes partitus* [13] are common.

The structure of coral reef fish communities could be influenced by natural and anthropogenic sources. The climate change may affect small sedentary fish more than large species [71]. However, the former might allow faster adaptation to new environmental conditions [72]. Given the general deterioration of reefs, a lower abundance is expected, not only of large carnivores but also of small specialist fishes [50].

The fishing pressure over fishes in coral reefs of the Gulf of Mexico and the Mexican Caribbean has produced changes in the richness and abundance of fishes [16, 50, 73]. For example, the fishing of groupers in the Mexican Caribbean caused the disappearance of the aggregation of Nassau grouper off Mahahual [74]. On the other hand, the scarcity of commercial fish species (e.g. snappers and groupers), become the parrotfish as a target group of spear-gun fishing as a result of the increasing of tourism along this area [73]. In the southern Mexican Caribbean there are a decreasing of coral reef fishes that is more evident in the large piscivores [50].

The anthropogenic disturbances, the tourism and river discharges are related to high nutrient levels on reef systems [21, 75], and could be linked to observed seagrass and hard coral cover loss over the last decades as in the Gulf of Mexico [21, 76] as in the Mexican Caribbean [77, 78]. The sediments and nutrients in coral reefs increase the turbidity and modify the richness and abundance of herbivorous fishes [66, 79].

5. The fisheries

An examination of exploited fish stocks of the southern Mexico and the Mexican Caribbean suggests that the main fisheries are composed by transient species, whose distribution extends to the warm-temperate region of the East and South Eastern USA; on the south side, fish fauna share components of Caribbean species whose distribution range up to the Brazilian coast in many cases. The main components of the exploited stocks are species dwelling the shelf grounds, and a smaller component is based on coral reef dwelling species. The general perception is that overfishing is a major concern for many reef-fish populations, and this activity may be one of the most important activities contributing to degradation of coral reefs in the southern Gulf of Mexico [17] and the Mexican Caribbean [16]. In the reefs of Veracruz, around 50 out of the 550 reef-fish recorded, are often used for fisheries. Some of them are caught directly on the coral reefs using hooks and harpoons [80, 81].

An overview of the fisheries of the Gulf of México [82] allows to conclude that the fisheries of the south and north Gulf display rather independent trends, being the Gulf menhaden (*Brevoortia patronus*) on the northern Gulf, the species ruling this trend, and representing more than 90% of the whole catch. In the southern Gulf the whole catch is more stable over time than the one of the northern Gulf and the Gulf menhaden does not appear in catch records, whose volume ranges between 50,000 and 100,000 t in the northern Gulf. On both sides, at the north and south Gulf, mullets became more abundant over the recent years, which together with shrimp and crabs as scavengers, suggest a probable increase of debris, caused by the intensity of fisheries trawling of the shrimp grounds. Other important exploited stocks deserving to be mentioned are the Spanish mackerel (*Scomberomorus maculatus*) and the King mackerel (*Scomberomorus cavalla*), two migratory species up in the food web as predators, running along the coasts of the state of Veracruz, with catch volumes of +5200 and 2300 t per year respectively. These two species also are important for fisheries on the north Gulf.

The fish species associated with reef and rocky areas (Gulf of Mexico and Mexican Caribbean) are usually reported in a group called "escama (scale fish)". They include groupers (e.g. *Epinephelus morio*), snappers (e.g. *Lutjanus jocu*), porgies (e.g. *Calamus bajonado*), grunts (e.g. *Haemulon plumierii*), hogfish (*Lachnolaimus maximus*), and tilefish (*Lopholatilus chamaeleonticeps*) among others [83].

In the 1970s, large predatory fishes such as sharks, and groupers were among the most important catches at the Mexican Caribbean reefs. The reduction of their populations led to new target species, such as Parrotfish, Whitefish, Spotted snapper, Tilefish and Creolefish [83]. Some of these fish species have been overfished [84]. Data by Arias-González [16] revealed that the biomass of large, predatory fish (Serranidae, Lutjanidae, Carangidae, and Sphyraenidae) was lesser in non-protected areas of the Mexican Caribbean.

In regard to the impact of climate of fish stocks, there is evidence [85] of a strong influence of climate indices, in particular the Southern Oscillation Index and the North Atlantic Oscillation Index on the catch of 66 species over historical records, finding that climate plays a significant role, in particular short-lived species. Some species respond with positive and others with negative sign respecting to the variability of these signs. This indicates the strong influence of climate. In some cases, trends help suggesting the most likely expectations of the catch in the near future, whilst others with declining trends make it difficult to forecast the effects of fishing intensity or other human impacts.

6. Conclusions

Species diversity of the Gulf of Mexico despite it does not show any evidence of being limiting for the Caribbean species, it somehow constrains the entrance to all typical Caribbean forms and in many zoological groups there are some species that are not found inside the Gulf and is known that some of these penetrate a far as the Campeche Bank reefs, or as occurs in some cases, they display very low abundances, like the Red grouper (*E. morio*), the Nassau grouper (*Epinephelus striatus*), the Gag (*Mycteroperca microlepis*). By contrast, there are a few fish species which are endemic to the Gulf of Mexico, like the Black snook (*Centropomus poeyi*), and Jarocho goby (*Elacatinus jarocho*), to just mention a few. After these examples, we can state that the Gulf of México contains some characteristics restricting the penetration of some Caribbean forms. It is pertinent to mention that in this chapter, mixohaline species associated to the brackish-waters are not considered.

In regard to the exploited fish stocks, two explanations are given respecting to significant changes of their biomasses, one is attributable to fishing intensity, and the other is to climatic variability. We consider that under certain circumstances, both factors may be responsible for these changes, especially if we look into particular fish stocks. Environmental variability is responsible of sudden changes in the biomass of short-lived species like sardine and related life forms, enhancing sudden increase or dramatic reductions [86] affecting productivity and the carrying capacity of the habitat. In contrast, long-lived stocks usually are able to support the effects of north winds occurring in winter, excepting those cases causing mass mortalities, and their biomasses are more related to their life spans and carrying capacity. The catch in the south Gulf of Mexico is estimated in 0.11 t/km^2 [87], and it was found a positive correlation of one year delay between yield and Chlorophyll *a* concentration.

Coral reefs of the Mexican Caribbean as well as those of the south of Gulf of Mexico have drastically changed over the last decades [41]. Recent studies showed an increase of macroalgae coverage and a decrease of coral cover [88] impacting coralline assemblages like herbivorous fish [73] by the lost of habitats and fishing pressure [16, 84]. The effects of fishing were detected in individual fish size and there were more evident on the unprotected reefs [16], where the commercially important species belonging to Serranidae, Lutjanidae, Carangidae, and Sphyraenidae were recorded by these authors in small quantities.

Acknowledgements

The authors would like to acknowledge the Mexican Ministry of Public Education (Secretaría de Educación Pública) for funding the project "Bases para el Análisis y Síntesis de los Sistemas Costeros de Veracruz" as part of the project "Analysis and Synthesis of the Coastal Zone of Veracruz, Gulf of Mexico".

Conflict of interest

The authors declare no conflict of interest.

Author details

Carlos González-Gándara[1]* and Ernesto A. Chávez[2]

1 Coral Reef Laboratory, Faculty of Biological and Agricultural Sciences, Veracruz University, Tuxpan, Veracruz, México

2 Interdisciplinary Center of Marine Sciences, IPN. Av. IPN s/n Playa Palo de Sta. Rita, El Conchalito, La Paz, Mexico

*Address all correspondence to: cggandara@hotmail.com

References

[1] Moyle PB, Cech JJ. Fishes. Fishes, An introduction to Ichthyology. 5th ed. India: Pearson Prentice Hall; 2004.726p.

[2] Hildebrand HH, Chávez H, Compton H. Aporte al conocimiento de los peces del arrecife Alacranes, Yucatán (México). Ciencia. 1964; 23(3):107-134.

[3] Reséndez-Medina A. Peces colectados en el arrecife La Blanquilla, Veracruz, México. Anales del Instituto de Biología Universidad Nacional Autónoma de México, Serie Ciencias del Mar y Limnología. 1971; 42(1):7-30.

[4] Bean TH. Notes on fishes collected at Cozumel, Yucatan, by the U.S. Fish Commission, with descriptions of new species. Bulletin of the United States Fisheries Commission. 1891; 8:193-206

[5] Díaz-Ruiz S, Aguirre-León A, Gazca-Castro M. Biodiversidad de los Peces en el Sistema Arrecifal Mesoamericano, Caribe Mexicano. In: Rivera-Arriaga E, Sánchez-Gil P, Gutiérrez J, editors. Tópicos de Agenda para la Sostenibilidad de Costas y Mares Mexicanos. 1st ed. Campeche: Universidad Autónoma de Campeche; 2019. p. 117-129. DOI: 10.26359/epomex.0519

[6] Schmitter-Soto JJ, Vásquez-Yeomans L, Aguilar-Perera A, Curiel-Mondragón C, Caballero-Vázquez JA. Lista de peces marinos del Caribe mexicano. Anales del Instituto de Biología, Universidad Nacional Autónoma de México, Serie Zoología. 2000; 71(2):143-177.

[7] Schmitter-Soto JJ. Peces. In: Pozo C, editor. Riqueza biológica de Quintana Roo. Un análisis para su conservación. Tomo 2. 1st ed. México: Colegio de la Frontera Sur, Comisión Nacional para el Conocimiento y Uso de la Biodiversidad, Gobierno del Estado de Quintana Roo, Programa de Pequeñas Donaciones México; 2011. p. 227-232.

[8] Chávez EA, CR Beaver. Reef Fish. In: Tunnell Jr JW, Chávez EA, Withers K, editors. Coral reefs of the southern Gulf of Mexico.1st ed. Corpus Christi: Texas A&M University Press; 2007. p. 102-111.

[9] Pérez-España H, Ávila-Gutiérrez PS, Melo-Merino SM, Berumen-Solórzano P, Flores-Arévalo RR. Patrones interanuales e interarrecifales de las comunidades de peces, corales y equinodermos en el Sistema Arrecifal Veracruzano. In: Granados-Barba A, Ortiz-Lozano LD, Salas-Monreal D, González-Gándara C, editors. Aportes al conocimiento del Sistema Arrecifal Veracruzano. Hacia el corredor arrecifal del suroeste del Golfo de México. 1st ed. Campeche: Universidad Autónoma de Campeche; 2015a. p. 159-178.

[10] Díaz-Ruiz S, Aguirre León A. Diversidad e ictiofauna de los arrecifes del sur de Cozumel, Quintana Roo. In: Salazar-Vallejo SI, González-Vallejo NE, editors. Biodiversidad costera y marina de México. 1st ed. Chetumal, México: Comisión Nacional para el Conocimiento y Uso de la Biodiversidad; 1993. p. 817-832.

[11] Loreto RM, Lara M, Schmitter-Soto JJ. Coral reef fish assemblages at Banco Chinchorro Mexican Caribbean. Bulletin of Marine Science. 2003; 73(1):52-170.

[12] Garduño M, Chávez EA. Fish resource allocation in coral reefs of Yucatán Peninsula. In: Munawar M, Lawrence SG, Munawar IF, Malley DF, editors. Aquatic Ecosistems of Mexico: Status and Scope. 1st ed. Netherlands: Backhuys Publishers; 2000. p. 367-381.

[13] Núñez-Lara E, Arias González JE. The relationship between reef fish community structure and environmental variables in the southern Mexican Caribbean. Journal of Fish Biology. 1998; 53 (Supl. A):209-221.

[14] Núñez-Lara E, Arias-González JE, Legendre P. Spatial patterns of Yucatan reef fish communities: testing models using a multi-scale survey design. Journal of Experimental Marine Biology and Ecology. 2005; 324:157-169.

[15] Rodríguez-Zaragoza FR, Arias-González EA. Additive Diversity Partitioning of Reef Fishes across Multiple Spatial Scales. Caribbean Journal of Science. 2008; 44 (1):90-101.

[16] Arias-González JE. Trophic models of protected and unprotected coral reef ecosystems in the South of the Mexican Caribbean. Journal of Fish Biology. 1998; 53 (Supl. A):236-255.

[17] Beaver CR, Chávez EA. Reef Fisheries. In: Tunnell Jr JW, Chávez EA, Withers K, editors. Coral reefs of the southern Gulf of Mexico.1st ed. Corpus Christi: Texas A&M University Press; 2007. p. 112-118.

[18] Liddell D. 2007. Origin and Geology. In: Tunnell Jr JW, Chávez EA, Withers K, editors. Coral reefs of the southern Gulf of Mexico.1st ed. Corpus Christi: Texas A&M University Press; 2007. p. 23-33.

[19] Ortiz-Lozano L, Colmenares-Campos C, Gutiérrez-Velázquez AL. Arrecifes sumergidos y su relevancia para el corredor arrecifal del suroeste del Golfo de México. In: Granados-Barba A, Ortiz-Lozano L, González-Gándara C, Salas-Monreal D, editors. Estudios científicos en el corredor arrecifal del suroeste del Golfo de México Campeche: Universidad Autónoma de Campeche; 2019. p. 1-36. DOI: 10.26359/epomex0319

[20] Ortiz-Lozano LD, Pérez-España H, Granados-Barba A, González-Gándara C, Gutiérrez-Velázquez A, Martos J. The Reef Corridor of the Southwest Gulf of Mexico: Challenges for its management and conservation. Ocean and Coastal Management. 2013; 86:22-32.

[21] Horta-Puga G, Tello-Musi JL, Beltrán-Torres A, Carricart-Ganivet JP, Carriquiry JD, Villaescusa-Celaya J. Veracruz reef system: a hermatypic coral community thriving in a sedimentary terrigenous environment. In: Granados-Barba A, Ortiz-Lozano LD, Salas-Monreal D, González-Gándara C, editors. Aportes al conocimiento del Sistema Arrecifal Veracruzano. Hacia el corredor arrecifal del suroeste del Golfo de México.1st ed. Campeche: Universidad Autónoma de Campeche; 2015. p. 181-207.

[22] Pérez-Maqueo O, Muñoz-Villiers L, Vázquez G, Equihua ZE, León RP. Hidrología. In: Cruz-Angón A, editor. La biodiversidad en Veracruz estudio de estado. Xalapa, México: Comisión Nacional para el Conocimiento y Uso de la Biodiversidad, Gobierno de Veracruz, Universidad Veracruzana, Instituto de Ecología; 2011. p. 289-292.

[23] Lara M, Padilla C, García C, Espejel JJ. Coral reefs of Veracruz, Mexico, I: Zonation and community. In: Proceedings of the 7th International Coral Reef Symposium; 22-27 June 1992; Mangilao, Guam: ICRS; 1992. p. 535-544.

[24] Jordan-Dahlgren E, Rodríguez-Martínez RE. The Atlantic coral reefs of México. In: Cortés J, editor. Latin American Coral Reefs, 1st ed. Amsterdam: Elsevier Science, 2003. p. 131-158.

[25] Chávez EA, Tunell Jr JW, Withers K. Reef Zonation and Ecology: Veracruz Shelf and Campeche Bank. In: Tunnell Jr JW, Chávez EA, Withers K, editors. Coral reefs of the southern Gulf of Mexico.1st ed. Corpus Christi: Texas A&M University Press; 2007. p. 41-67.

[26] González-Gándara C, Salas-Pérez JJ. Caracterización del sustrato bentónico del arrecife Tuxpan, Veracruz, México, con énfasis en la comunidad de corales escleractinios. Revista de Biología Marina y Oceanografía. 2019; 54(3):259-271. DOI:10.22370/rmbo.2019.54.3.2009

[27] Wenger AS, Williamson DH, da Silva ET, Ceccarelli DM, Browne NK, Petus C, Devlin MJ. Effects of reduced water quality on coral reefs in and out of no-take marine reserves. Conservation Biology. 2015; 30(1):142-153. DOI: 10.1111/cobi.12576

[28] Rogers JS, Monismith SG, Koweek DA, Torres WI, Dunbar RB. Thermodynamics and hydrodynamics in the atoll reef system and their influence on coral cover. Limnology and Oceanography. 2016; 61:2191-2206. DOI: 10.1002/lno.10365

[29] De la Cruz-Francisco V, Bandala-Pérez AE. Esponjas y Cnidarios (Hydrozoa y Anthozoa) del arrecife Oro verde; cobertura bentónica y afinidad faunística con sistemas arrecifales de Veracruz, México. CICIMAR Oceánides. 2016; 31(2):45-64.

[30] González-Gándara C. La comunidad de corales escleractinios del sistema arrecifal Palo Seco, Veracruz, México. In: Pérez-Morales A, Ake-Castillo JA, Álvarez-González CA, Poot-Delgado CA, editors. Investigaciones marinas en el Golfo de México y Mar Caribe Mexicano. 1st ed. Colima: Universidad de Colima; 2020. p.186-204.

[31] Martos FJ. Evaluación de la salud de los corales del arrecife Blake, Cazones, Veracruz, México [Thesis]. Xalapa, Veracruz, México: Universidad Veracruzana; 2010.

[32] Pérez-España H, Bello-Pineda J, Ávila-Santiago P, Berumen-Solorzano P, Melo-Merino S, Flores-Arévalo RR, Santander-Monsalvo J, Gómez-Villada RS, Román-Vives MA, Ruiz-Lozano C, Alvardo-Tejeda M, Roldán-Ubando AA, Cabrera-Valenzuela A. Monitoreo del Parque Nacional Sistema Arrecifal Veracruzano: Segunda Etapa. México: Informe final SNIB-CONABIO, proyecto GM004. 2015. p. 1-97.

[33] González-Gándara C. Composition, distribution and relative abundance of fishes linked to fringing reefs of southern Veracruz, Gulf of Mexico. GSC Biological and Pharmaceutical Sciences. 2020; 12(02):217-228. DOI:10.30574/gscbps.2020.12.2.0268

[34] Bonet F. Biogeología subsuperficial del arrecife Alacranes, Yucatán. Boletín del Instituto de Geología, Universidad Nacional Autónoma de México. 1967; 80:1-192.

[35] Jordan-Dahlgren E. El Ecosistema Arrecifal Coralino del Atlántico Mexicano. Revista Sociedad Mexicana de Historia Natural. 1993; 44:157-175.

[36] Logan BW. Coral reefs and Banks, Yucatan shelf, Mexico. In: Logan BW, Harding JL, Ahr WM, Williams JD, Snead RG, editors. Carbonate sediments and reefs, Yucatan Shelf, Mexico. American Association of Petroleum Geologists Memoir; Tulsa, Oklahoma: AAPG; 1969. p. 129-198

[37] Zarco-Perelló S, Mascaró M, Garza-Pérez R, Simoes N. Topography and coral community of the Sisal Reefs, Campeche Bank, Yucatán, México. Hidrobiológica. 2013; 23(1):28-41.

[38] Robertson DR, Domínguez-Domínguez O, Lopez Aroyo YM, Moreno Mendoza R, Simoes N. Reef-associated fishes from the offshore reefs of western Campeche Bank, Mexico, with a discussion of mangroves and seagrass beds as nursery habitats. ZooKeys. 2019; 843:71-115. DOI: 10.3897/zookeys.843.33873

[39] Carrillo L, Horta-Puga G, Carricart-Ganivet JP. Climate and Oceanography. In: Tunnell Jr JW, Chávez EA, Withers K, editors. Coral reefs of the southern Gulf of Mexico.1st ed. Corpus Christi: Texas A&M University Press; 2007. p. 34-40.

[40] Johnston MW, Bernard AM. A bank divided: quantifying a spatial

and temporal connectivity break between the Campeche Bank and the northeastern Gulf of Mexico. Marine Biology. 2017; 164:12. DOI 10.1007/s00227-016-3038-0

[41] Rioja-Nieto R, Álvarez-Filip L. Coral reef systems of the Mexican Caribbean: Status, recent trends and conservation. Marine Pollution Bulletin. 2019; 140:616-625.

[42] Reyes-Bonilla H, Jordan-Dahlgren E. Caribbean Coral Reefs: Past, Present, and Insights into the Future. In: Rossi S, Bramanti L, Gori A, Orejas C, editors. Marine Animal Forests. 1st ed. Cham: Springer International Publishing; 2017. p. 31-72. DOI: 10.1007/978-3-319-17001-5_2-1

[43] Comisión Nacional de Áreas Naturales Protegidas. Programa de Manejo Parque Marino Nacional Arrecifes de Cozumel.1st ed. México: Comisión Nacional de Áreas Naturales Protegidas, Secretaría de Medio Ambiente y Recursos Naturales; 1998. 164p.

[44] Comisión Nacional de Áreas Naturales Protegidas. Programa de Manejo de la Reserva de la Biosfera Banco Chinchorro. 1st ed. México: Comisión Nacional de Áreas Naturales Protegidas, Secretaría de Medio Ambiente y Recursos Naturales; 2000. 192p.

[45] Comisión Nacional de Áreas Naturales Protegidas. Programa de Manejo del Parque Nacional Arrecife de Puerto Morelos. 1st ed. México: Comisión Nacional de Áreas Naturales Protegidas, Secretaría de Medio Ambiente y Recursos Naturales; 2000. 224p.

[46] Comisión Nacional de Áreas Naturales Protegidas. Programa de Manejo Parque Nacional Arrecifes de Xcalak. 1st ed. México: Comisión Nacional de Áreas Naturales Protegidas, Secretaría de Medio Ambiente y Recursos Naturales; 2004. 162p.

[47] Comisión Nacional de Áreas Naturales Protegidas. Programa de Conservación y Manejo Parque Nacional Arrecife Alacranes.1st ed. México: Comisión Nacional de Áreas Naturales Protegidas, Secretaría de Medio Ambiente y Recursos Naturales; 2006. 168p.

[48] Comisión Nacional de Áreas Naturales Protegidas. Programa de Manejo Complejo Sian Ka'an: Reserva de la Biosfera Sian Ka'an, Área de Protección de Flora y Fauna Uaymil y Reserva de la Biosfera Arrecifes de Sian Ka'an. 1st ed. México: Comisión Nacional de Áreas Naturales Protegidas, Secretaría de Medio Ambiente y Recursos Naturales; 2014. 481p.

[49] Lorda JF, Athíe G, V Camacho I, Daessle LW, Molina O. The relationship between zooplankton distribution and hydrography in oceanic waters of the Southern Gulf of Mexico. Journal of Marine Systems. 2019; 192:28-41. DOI: 10.1016/j.jmarsys.2018.12.009

[50] Schmitter-Soto JJ, Aguilar-Perera A, Cruz-Martínez A, Herrera-Pavón RL, Morales-Aranda AA, Cobián-Rojas D. Interdecadal trends in composition, density, size, and mean trophic level of fish species and guilds before and after coastal development in the Mexican Caribbean. Biodiversity and Conservation. 2018; 27: 459-474. DOI: 10.1007/s10531-017-1446-1

[51] González-Gándara C, Lozano-Vilano ML, De la Cruz-Francisco V, Domínguez Barradas C. Peces del sistema arrecifal Lobos-Tuxpan, Veracruz, México. Universidad y Ciencia. 2013; 29(2): 191-208.

[52] González-Gándara C. Peces del arrecife Blake, Veracruz, México:

Inventario, distribución y afinidades zoogeográficas. Ecosistemas y Recursos Agropecuarios. 2015; 2(4): 87-97.

[53] Del Moral-Flores LF, Tello-Musi JL, Reyes-Bonilla H, Pérez-España H, Martínez-Pérez JA, Horta-Puga G, Velazco-Mendoza LA, Álvarez del Castillo-Cárdenas PA. Lista sistemática y afinidades zoogeográficas de la ictiofauna del Sistema Arrecifal Veracruzano, México. Revista Mexicana de Biodiversidad. 2013; 84: 825-846. DOI: 10.7550/rmb.34912

[54] Robertson DR, Pérez-España H, Domínguez-Domínguez O, Estapé CJ , Estapé AM. An update to the inventory of shore-fishes from the Parque Nacional Sistema Arrecifal Veracruzano, Veracruz, México. ZooKeys. 2019; 882: 127-157. DOI: 10.3897/zookeys.882.38449

[55] González-Gándara C, Arias-González JE. Lista actualizada de los peces del Arrecife Alacranes, Yucatán, México. Anales del Instituto de Biología. Universidad Nacional Autónoma de México. 2001; 72(2):245-258.

[56] Zarco Perelló S, Moreno Mendoza R, Simões N. Checklist of Fishes from Madagascar Reef, Campeche Bank, México Biodiversity Data Journal. 2014; 2: e1100. DOI: 10.3897/BDJ.2.e1100

[57] Briggs JC. Marine centres of origin as evolutionary engines. Journal of Biogeography. 2003; 30:1-18

[58] González-Gándara C. Las comunidades de peces del Arrecife Alacranes, Yucatán, México: Variaciones espacio-temporales. [PHD Thesis]. Mérida, Yucatán, México: Centro de Investigaciones y de Estudios Avanzados. Instituto Politécnico Nacional; 2001.

[59] Nava MGG. 2006. Comunidades de peces y sus relaciones con el hábitat en dos sistemas arrecifales del Golfo de México. [Master Thesis]. Mérida, Yucatán, México: Centro de Investigaciones y de Estudios Avanzados. Instituto Politécnico Nacional; 2006.

[60] Darling ES, Graham NAJ, Januchowski-Hartley FA, Nash KL, Pratchett MS, Wilson SK. Relationships between structural complexity, coral traits, and reef fish assemblages. Coral Reefs. 2017; DOI: 10.1007/s00338-017-1539-z

[61] Brandl SJ, Johansen JL, Casey JM, Tornabene L, Morais RA, Burt JA. Extreme environmental conditions reduce coral reef fish biodiversity and productivity. Nature Communications. 2020; 11:3832. DOI: 10.1038/s41467-020-17731-2.

[62] De Souza CS, Mafalda-Junior PO. Large-Scale Spatial and Temporal Variability of Larval Fish Assemblages in the Tropical Atlantic Ocean. Anais da Academia Brasileira de Ciências. 2019; 91(1):e20170567. DOI: 10.1590/0001-3765201820170567.

[63] Cooper AM, MacDonald C, Roberts E, Bridge TCL. Variability in the functional composition of coral reef fish communities on submerged and emergent reefs in the central Great Barrier Reef, Australia. PLoS ONE. 2019; 14(5):e0216785. DOI: 10.1371/journal. pone.0216785

[64] Résendez-Medina A. Estudio de los peces de la Laguna de Términos, Campeche, México. II. Biótica. 2001; 6:345-430.

[65] Raz-Guzmán A, Huidobro L, Padilla V. An updated checklist and characterisation of the ichthyofauna (Elasmobranchii and Actinopterygii) of the Laguna de Tamiahua, Veracruz, Mexico. Acta Ichthyologica et Piscatoria. 2018; 48 (4):341-362. DOI: 10.3750/AIEP/02451

[66] Bejarano I, Appeldoorn R. Seawater turbidity and fish communities on coral reef of Puerto Rico. Marine Ecology Progress Series. 2013; 474:217-226. DOI: 10.3354/meps10051

[67] Gamboa-Pérez HC, Schmitter-Soto JJ. Distribution of cichlid fishes in the littoral of Lake Bacalar, Yucatan Peninsula. Environmental Biology of Fishes. 1999; 54:35-43.

[68] Aldana Moreno A, Montero Muñoz J, Aldana Aranda D. Variación espacio-temporal de la ictiofauna del Parque marino Xel-Há, Caribe mexicano y su relación con parámetros fisicoquímicos. Revista de Biología Tropical. 2016; 64 (4):1353-1367. DOI:10.15517/rbt.v64i4.22703

[69] Nagelkerken I, Kleijnen S, Klop T, van den Brand RACJ, Cocheret de la Morinière E, van der Velde G. Dependence of Caribbean reef fishes on mangroves and seagrass beds as nursery habitats: a comparison of fish faunas between bays with and without mangroves/seagrass beds. Marine Ecology Progress Series. 2001; 214:225-235.

[70] Nagelkerken I, Huebert KB, Serafy JE, Grol MGG, Dorenbosch M, Bradshaw JA Highly localized replenishment of coral reef fish populations near nursery habitats. Marine Ecology Progress Series. 2017; 568: 137-150. DOI: 10.3354/meps12062

[71] Ottimofiore E, Albouy C, Leprieur F, Descombes P, Kulbicki M, Mouillot D, Parravicini V, Pellissier L. Responses of coral reef fishes to past climate changes are related to life-history traits. Ecology and evolution. 2017; 7(6): 1996-2005. DOI: 10.1002/ece3.2800

[72] Munday PL, Jones GP, Pratchett MS, Williams AJ. Climate change and the future for coral reef fishes. Fish and Fisheries. 2008; 9(3): 261-285.

[73] Hernández-Landa RC, Acosta-González G, Nuñez-Lara E, Arias-González EA. Spatial distribution of surgeonfish and parrotfish in the north sector of the Mesoamerican Barrier Reef System. Marine Ecology. 2014; 36(3):432-446. DOI: 10.1111/maec.12152

[74] Aguilar-Perera A. Disappearance of a Nassau grouper spawning aggregation off the southern Mexican Caribbean coast. Marine Ecology Progress Series. 2006; 327: 289-296. DOI: 10.3354/meps327289

[75] Baker DM, Rodríguez-Martínez RE, Fogel ML. Tourism's nitrogen footprint on a Mesoamerican coral reef. Coral Reefs. 2013; 32: 691-699. DOI: 10.1007/s00338-013-1040-2

[76] Ramírez-García P, Hernández-Martínez O, Pedraza K, Quiroz A. State of *Thalassia testudinum* Banks ex König meadows in the Veracruz Reef System, Veracruz, México. Aquatic Botany. 2008; 88: 17-26. DOI: 10.1016/j.aquabot.2007.08.003

[77] van Tussenbroek BI, Cortés J, Collin R, Fonseca AC, Gayle PMH, Guzmán HM, Jácome GE, Juman R, Koltes KH, Oxenford HA, Rodríguez-Ramirez A, Samper-Villarreal J, Smith RS, Tschirky JJ, Weil E. Caribbean-Wide, Long-Term Study of Seagrass Beds Reveals Local Variations, Shifts in Community Structure and Occasional Collapse. PLoS ONE. 2014; 9(3): e90600. DOI: 10.1371/journal.pone.0090600

[78] Arias-González JE, Fung T, Seymour RM, Garza-Pérez JR, Acosta-González G, Bozec YM, Johnson CR. A coral-algal phase shift in Mesoamerica not driven by changes in herbivorous fish abundance. PLoS ONE. 2017; 12(4): e0174855. DOI: 10.1371/journal.pone.0174855

[79] Wolanski E, Richmond R, McCook L. A model of the effects

of land-based, human activities on the health of coral reefs in the Great Barrier Reef and in Fouha Bay, Guam, Micronesia. Journal of Marine Systems. 2004; 46: 133-144.

[80] Vargas-Hernández JM, Jiménez-Badillo ML, Arenas-Fuentes V. El sistema arrecifal Veracruzano y las pesquerías asociadas. In: Amaya PC, Quiroga-Brahms C, Luna CD, Castellanos DF, Contreras CM, Silva-López G, editors. La Pesca en Veracruz y Sus Perspectivas de Desarrollo, 1st ed. México: Secretaría de Agricultura, Ganadería, Desarrollo Rural, Pesca y Alimentación, Universidad Veracruzana; 2002. p. 13-16.

[81] Jiménez-Badillo L, Pérez-España H, Vargas-Hernández JM, Cortés-Salinas JC, Flores-Pineda PA. Catálogo de Especies y Artes de Pesca del Parque Nacional Sistema Arrecifal Veracruzano. 1st ed. Xalapa: Comisión Nacional para el Conocimiento y Uso de la Biodiversidad, Universidad Veracruzana; 2006. 182 p.

[82] Chávez EA, Chávez-Hidalgo A. Fisheries production of the Gulf of Mexico. In: Proceedings of 65th Gulf and Caribbean Fisheries Institute; 5-9 November 2012; Santa Marta Colombia. GCFI; 2013. p. 87-94.

[83] Fernández JI, Álvarez-Torres P, Arreguín-Sánchez F, López-Lemus LG, Ponce G, Díaz-de-León A, Arcos-Huitrón E, del Monte-Luna P. Coastal fisheries of Mexico. In: Salas S, Chuenpagdee R, Charles A, Seijo JC, editors. Coastal fisheries of Latin America and the Caribbean. FAO Fisheries and Aquaculture Technical Paper. No. 544. Rome: FAO; 2011. p. 231-284.

[84] Jackson JBC, Donovan MK, Cramer KL, Lam VV, editors. Status and Trends of Caribbean Coral Reefs: 1970-2012. 1st ed. Gland, Switzerland: Global Coral Reef Monitoring Network c/o International Union for the Conservation of Nature; 2014. 304p.

[85] Chávez EA, Chávez-Hidalgo A, Castro-Ortiz. Effect of Climate Variability on Fish Stocks of the Northern Gulf of Mexico. Proceedings of the 68th Gulf and Caribbean Fisheries Institute; 9-13 November 2015; Panamá: 2016. p. 456-466.

[86] Chávez EA, Chávez-Hidalgo A. Biomass from the sea. In: Matovic MD editor. Biomass Now-Sustainable Growth and Use. London: Intechopen; 2013. p. 511-522. DOI: 10.5772/54520

[87] Chávez EA, Chávez-Hidalgo A, Pérez-Lezama E. Fisheries potential of the Gulf of Mexico. Japan Journal of Medical Science. 2020; 1(2):16-19.

[88] Martínez-Rendis A, Acosta-González G, Hernández-Stefanoni LJ, Arias-González JE. Quantifying the reefscape transformation of a coastal Caribbean coral reef during a phase shift and the associated coastal landscape. Marine Ecology. 2015; 37:697-710. DOI: 10.1111/maec.12334

Chapter 8

More Urbanization, Fewer Bats: The Importance of Forest Conservation in Honduras

Manfredo A. Turcios-Casco, Richard K. LaVal, Marcio Martínez and Hefer D. Ávila-Palma

Abstract

Urbanization is a phenomenon that results in fragmentation and eventual destruction of forests. Suburbanization is a subset of that same phenomenon in which fragmentation has resulted in the retention of small patches of the original forest and surviving old growth trees. Alternatively, the area surrounding the central city had been cleared for agricultural use and the suburban residents have planted many trees in parks and private property. This fragmentation will of course affect many species of bats, including species of the family Phyllostomidae. In this work, we estimate and compare the diversity of phyllostomid bats in three landscapes in Honduras: forests, suburban, and urban areas, from 2015 to 2018. Concurrently, we compared bat activity patterns based on the hour and percentage of moonlight at the time they were captured, and we compared external measurements, forearm and ear length. Urban areas are the least diverse and exhibited the lowest abundance. The forearm and ear length were significantly different only between forests and urban areas. The degree of lunar phobia also differed among those landscapes, but the time of capture did not differ. This is the first attempt to describe the activity patterns of phyllostomids in these studied areas and the effect of urbanization on Honduran bats. As expected, we found that from forests to cities, the diversity and abundance of phyllostomids decreased. However, there are many gaps in our knowledge of how totally or partially urbanized areas are affecting phyllostomid bats in Honduras.

Keywords: Chiroptera, CU-UNAH, Honduras, phyllostomid bats, Río Plátano Biosphere Reserve, Sabanagrande, suburbanization, taxonomic diversity

1. Introduction

1.1 Urbanization, suburbanization, and bats

It is generally agreed that urbanization has had a major negative effect on populations and diversity of native plants and animals. Nevertheless, humans plant a wide variety of vegetation in their urban and suburban areas, thus initiating the food chain on which most animals depend. Although diversity is clearly greatly reduced in the urban setting, at the population level the effect has been extremely heterogeneous across the animal kingdom. Bats (Chiroptera) are an extremely diverse group, with more than 1400 species worldwide, living in almost all habitats. The reaction of bats to urban environments was recently reviewed [1]. Jung and Caragh [1] determined that the behavioral and/or morphological traits at the level of individual species determine species' adaptability to urban areas. Further, they determined that the driving factors for species adaptability to urban areas may be regionally divergent.

As Jung and Caragh [1] point out, bats are found in all cities over the world. Of the approximately 20 families of bats, only two tend to avoid cities entirely, the Rhinolophidae and Mormoopidae, whereas a heterogeneous reaction at the species level is typical of the other families.

Urban habitats have both potential disadvantages and advantages from the perspective of bats. Clearly cities are high in noise, light, and chemical pollution compared to natural habitats. Light pollution may be an especially difficult factor to which bats must adapt. Depending on roosting requirements, cities may provide abundant roost sites, such as buildings, or not, for example for bats that roost in vegetation. Often drinking water and food supplies are enhanced by the human residents of the city, again depending on the bats' specific requirements. Since tree cover in cities averages less than 30% [2], bats adapted to forests may not do well, whereas in grassland and savannah areas, bats may find the tree cover advantageous [3].

Although there have been many studies of urban bats, these have been concentrated in temperate North America and Europe, and focus mainly on bats of the family Vespertilionidae, e.g., Dixon [4], Hale et al. [5], Pearce and Walters [6]. Bat activity and diversity seem to be highest in older suburban areas and parks and decrease towards the center of cities where there is little vegetation. It is clear to us that the change from urban to suburban to rural is a continuum; therefore, it is not productive to divide this continuum into discrete units except very generally as we are doing here for comparative purposes. The differences between urban and suburban can, for example be exemplified by such physical differences as tree density, percent paved area, building size and density, etc. These variables change in a predictable way as we pass through the continuum. Thus, when we approach 0% paved area, very low building density, and/or 100% tree density, we have reached the end of the continuum and are in forest or agricultural zones.

Although a few species do very well in cities, as for example, the huge colony of *Tadarida brasiliensis* in downtown Austin, Texas, U.S.A. [7], the majority of species that occur in a given area are rare or absent from urban areas [1]. Norberg and Rayner [8], pointed out that bat species with high wing loadings and aspect ratios, and thus presumed to forage in open areas, seem to be the most abundant and diverse in cities. Several studies [9–11] show that in general, foraging activity of bats seems to be higher in rural and forested areas than in urban areas.

A threat to bat populations, clearly related to urbanization, is mortality on highways. This problem has been but rarely studied, mostly in the temperate zone (e.g., [12]) but clearly exists. Recently a study in Brazil demonstrated that significant bat mortality is occurring on Brazilian highways as well [13].

There are relatively few studies of bats in urban areas of the Neotropics. Jung and Kalko [14] in Panama, using audio recordings, report decreased diversity and abundance in the urban setting, compared to the high diversity in forests in that country. They also note that in the city, most of the bats are high flying species, primarily of the Molossidae. In Costa Rica one of us [15] found the same trend with audio recordings in a large metropolis and in a smaller provincial capital. Jung and Kalko [16], recording in a small city adjacent to forest, found higher diversity than in large urban centers, but noted that some species that were abundant in the nearby forest were never recorded in town. For our purposes, we may think of this town as a suburban area. In Costa Rica, in a large urban center, but recording only in city parks, the number of bat passes was much greater in the larger parks than in the smaller parks [17]. Overall, those authors found considerably less activity than we consistently found in non-suburban settings [14]. The urban bats identified from the calls were all from the families Vespertilionidae and Molossidae. In another Costa Rican study the author mist-netted in parks in the city, finding a relatively small number (for the netting effort) of bats of the family Phyllostomidae, all of which were very common species that eat fruit and/or nectar [18].

1.2 Bat diversity in Honduras

Because of its location on the relatively narrow isthmus of Central America, connecting North and South America, Honduras is home to species typical of South America, others typical of temperate North America, and some that are endemic to Central America and southern Mexico. According to a recent review, [19] there are 113 species of bats currently known from Honduras, and we expect several more species will be added in the future. In Central America, only Costa Rica has more species, with 120 listed [20]. As detailed in the paper cited [19], the bats belong to seven different families, the Emballonuridae (9 species), the Phyllostomidae (59 species), the Mormoopidae (5 species), the Noctilionidae (2 species), The Thyropteridae (1 species), the Natalidae (2 species), the Molossidae (17 species), and the Vespertilionidae (18 species).

These species include frugivores, nectarivores, insectivores, sanguinivores, carnivores, and omnivores, and occupy many essential ecological niches in Honduras, dispersing seeds, pollinating flowers, and controlling insect numbers, among others.

1.3 Objectives and hypothesis

We lack information demonstrating how urbanization is affecting the diversity of bats in Honduras, Central America. We hypothesize that, on a continuum from forests to cities, the diversity of New World leaf-nosed bats (Phyllostomidae) will be significantly reduced. Therefore, this works aims to estimate and compare the diversity of phyllostomid bats in three landscapes in Honduras: forests, suburban, and urban areas; to determine if the forearm and ear length has any significant effect on species composition of bat assemblages in the three landscapes; and to describe the activity patterns of those assemblages.

2. Materials and methods

2.1 Concepts

When defining "urbanization" and "suburbanization", which are processes that are closely related and linked along a continuum, we follow Tammaru et al. [21]. We will consider suburbanization as the expansion of suburbs by the increase of its population from the migration of residents of the central city [21, 22]. Thereby, we will refer to the Ciudad-Universitaria of the Universidad Nacional Autónoma de Honduras (CU-UNAH) as the "urban area" in this manuscript. The same authors described suburbanization is the redistribution of a population away from central cities and into suburbs. In this work, we are referring to Sabanagrande as the "suburban area". All other studied areas in this work are considered as "forests" located in the Río Plátano Biosphere Reserve. See below for the description of each studied area.

2.2 Studied areas

All the coordinates are given in **Table 1** and represented in **Figure 1**, and each site is described below:

2.2.1 CU-UNAH

Surveys were carried out at the National University Campus' Botanical Garden in Tegucigalpa, capital city of Honduras, in the Department of Francisco Morazán (**Table 1**). The ecosystem is a tropical dry forest [23]. The species of Fabaceae, Myrtaceae, and Asteraceae are the most common, including *Muntingia calabura*, *Byrsonima crassifolia*, *Lonchocarpus sanctuarii* and cultivated plants, such as *Hibiscus rosa-sinensis*, *Russelia equisetiformis* and *Psidium guajava* [24].

2.2.2 Sabanagrande

We studied a tropical moist forest [23] located in the central region of Honduras in the Department of Francisco Morazán, municipality of Sabanagrande (**Table 1**). The vegetation included *Pinus oocarpa*, *P. maximinoi*, *Quercus oleoides*, and plants of the genus *Miconia* (Melastomataceae), *Curatella* (Dilleniaceae), *Psidium* (Myrtaceae), *Calliandra* (Fabaceae), and *Ficus* (Moraceae). Extensive livestock and crops of *Zea mays* (Poaceae) can also be found near the studied areas [25].

2.2.3 RPBR

The RPBR, including La Moskitia, is located within the departments of Gracias a Dios, Olancho, and Colón. Based on Holdridge [23], the life zone represented is tropical wet forest. The RPBR is the only site in Honduras declared as world Heritage. Some plant species associated with the study area are *Swietenia macrophylla* (Meliaceae), *Cedrela odorata* (Meliaceae), *Cordia alliodora* (Meliaceae), *Chamaedorea tepejilote* (Arecaceae), *Geonoma congesta* (Arecaceae), *Cecropia obtusifolia* (Urticaceae), *Psychotria poeppigiana* (Rubiaceae), and *Sloanea picapica* (Eleocarpaceae) [26].

2.3 Taxonomy, mist-netting, and ethical guidelines

We followed the most recent taxonomic checklist of the bats that occur in Honduras [19]. All the bats were captured using mist-nets of 12.5 x 2.5 m with a mesh of 35 mm. We followed Kunz and Kurta [27] for positioning the mist nets according to the vegetation, landscape, bodies and water and phenophases (fruits and flowers) of the plants. Bats were identified and measured (FA = forearm length; E = ear length; BH = body height) using taxonomic keys of Timm, LaVal and Rodriguez [28] Medina-Fitoria [29], and Mora [30]. We quantified the sampling effort by multiplying the area of all of the mist-nets that were opened during each night by the number of hours that remained open [31] in which a total of 47,686.8 $m^{2}*h$ was accumulated. All the bats were handled according to the guidelines for the use of wild mammals in research and education [32].

	Departament	Locality	Municipality	Latitude	Longitude	Elevation (m asl)	Landscape
1	Francisco Morazán	Carboneras	Sabanagrande	13.794	−87.248	985	Suburban
2	Gracias a Dios	Ciudad Blanca 1	Brus Laguna	15.246	−84.969	250	Forest
3	Gracias a Dios	Ciudad Blanca 2	Brus Laguna	15.246	−84.972	214	Forest
4	Gracias a Dios	Ciudad Blanca 3	Brus Laguna	15.245	−84.96	245	Forest
5	Gracias a Dios	Ciudad Blanca 4	Brus Laguna	15.245	−84.969	225	Forest
6	Gracias a Dios	Ciudad Blanca 5	Brus Laguna	15.248	−84.968	223	Forest
7	Gracias a Dios	Ciudad Blanca 6	Brus Laguna	15.245	−84.965	204	Forest
8	Gracias a Dios	Ciudad Blanca 7	Brus Laguna	15.251	−84.974	239	Forest
9	Gracias a Dios	Ciudad Blanca 8	Brus Laguna	15.244	−84.966	233	Forest
10	Gracias a Dios	Ciudad Blanca 9	Brus Laguna	15.241	−84.969	206	Forest
11	Francisco Morazán	El Ocotal	Sabanagrande	13.791	−87.314	976	Suburban
12	Francisco Morazán	La Finca "Divisadero"	Sabanagrande	14.561	−87.801	1105	Suburban
13	Francisco Morazán	La Tigra	Sabanagrande	13.800	−87.313	790	Suburban
14	Gracias a Dios	Las Marías Pesh 1	Juan Francisco Bulnes	15.680	−84.838	33	Forest
15	Gracias a Dios	Las Marías Pesh 2	Juan Francisco Bulnes	15.679	−84.846	50	Forest
16	Gracias a Dios	Las Marías Pesh 3	Juan Francisco Bulnes	15.676	−84.851	28	Forest
17	Gracias a Dios	Las Marías Pesh 4	Juan Francisco Bulnes	15.676	−84.843	33	Forest
18	Gracias a Dios	Pico Dama 1	Juan Francisco Bulnes	15.695	−84.915	373	Forest
19	Gracias a Dios	Pico Dama 2	Juan Francisco Bulnes	15.695	−84.915	360	Forest
20	Gracias a Dios	Pico Dama 3	Juan Francisco Bulnes	15.692	−84.915	394	Forest
21	Gracias a Dios	Pico Dama 4	Juan Francisco Bulnes	15.695	−84.917	433	Forest
22	Gracias a Dios	Pico Dama 5	Juan Francisco Bulnes	15.694	−84.915	383	Forest
23	Francisco Morazán	UNAH (CC)	Distrito Central	14.008	−87.165	1073	Urban
24	Francisco Morazán	UNAH (JB)	Distrito Central	14.087	−87.166	1050	Urban

	Departament	Locality	Municipality	Latitude	Longitude	Elevation (m asl)	Landscape
25	Francisco Morazán	UNAH (Lagunas)	Distrito Central	14.086	−87.160	1050	Urban
26	Francisco Morazán	UNAH (Polideportivo)	Distrito Central	14.086	−87.169	1062	Urban
27	Gracias a Dios	Waikna Tara	Juan Francisco Bulnes	15.660	−84.893	44	Forest

Table 1.
Description of all the 27 localities that were studied and were classified among urban, suburban, and forests.

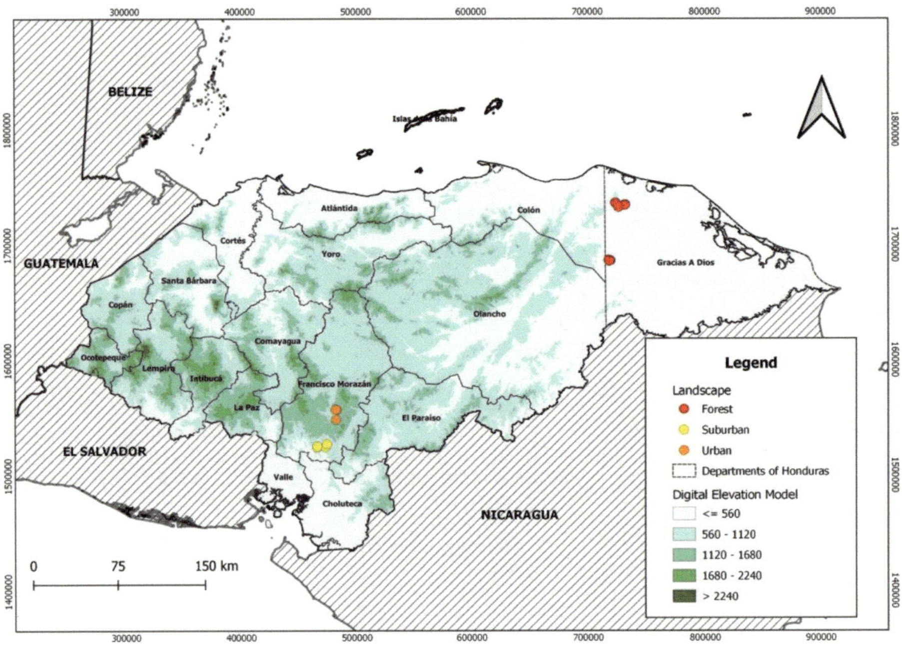

Figure 1.
Forest, suburban, and urban areas used in this study.

2.4 Diversity, landscape, and activity patterns analyses

The diversity of each landscape was measured using the Alpha Diversity Index (following Jost [33] and Moreno et al. [34]), and species richness was estimated with Chao 1. These analyses were based in the sampling effort of each site and the abundances of each species using EstimateSMac 910 with 100 randomizations [35, 36].

For the activity patterns analyses we used the abundances of all the species [37] and correlated them with the time and the percentage of the moon illumination in which bats were captured from each type of landscape. Moon illumination was taken for each date from the following website: https://www.moongiant.com/ [38]. We used the Shapiro–Wilk test to test for the normal distribution and Levene's test to test for the homogeneity of variances of the data. Considering that data was normally distributed, means of the forearm length, ear length, body height, time, and moon percentage were represented by the ANOVA (Analysis of Variance) analyses in **Table 2**. To compare means we performed posthoc Tukey tests at a

DOI: http://dx.doi.org/10.5772/intechopen.96493

Landscape	Individuals	Species	Sampling effort (m^2*h)	Chao 1 estimator	Alpha diversity index	Individuals per m^2*h	Species per m^2*h	FA (mm)	E (mm)	BH (mm)	Moon illumination (%)	Time
Forests	376	24	14,567.4	28.97	5.72	0.03	0.0016	74.68 (29.00–91.35)	30.75 (9.48–42.54)	53.29 (36.8–69.78)	65.26 (15.5–99.52)	5:30 PM – 5:20 AM
Suburban	169	17	18,839.4	19.24	4.71	0.01	0.0009	67.74 (33.01–69.46)	16.59 (7.67–17.83)	57.60 (45.9–69.30)	47.46 (0.24–94.43)	6:00 PM – 2:00 AM
Urban	143	7	14,280.0	7.5	1.54	0.01	0.0005	67.71 (31.10–72.00)	23.99 (9.82–28.34)	68.73 (55.96–81.50)	50.06 (0.06–100)	5:30 PM – 2:20 AM

Table 2.
Diversity data and means of the morphometrics and ecological data of the 688 bats studied.

confidence level of 95%. For all the analyses we considered significant differences when $p \leq 0.05$. Except for time, in which we use the Spearman's correlation coefficient to determine any relation between the landscapes and the time activity.

To determine taxonomic similarities between the landscapes (urban and suburban areas and forest) we performed multiple regressions of distance matrices [39]. In addition, to represent graphically the taxonomic composition in the distinct habitat types, we performed a NMDS (non-metric multi-dimensional scaling) analysis with two dimensions and plotted the NMDS axes against landscapes [40]. All analyses were performed in R Core Team [41] version 3.4.2, using the vegan [42] and ecodist [43].

3. Results

3.1 Alpha diversity in three landscapes

According to Chao 1 (**Table 2**), urban areas (percentage of how many species According to Chao 1, urban areas (percentage of how many species were recorded in parentheses) are the least diverse, as expected, because only 7.5 species are expected (93.3%) followed by suburban areas (87.2%) with 19.24, and then by forests with 28.97 (82.2%). Supporting Chao 1, the Alpha diversity index was highest in forests with 5.72, followed by suburban and urban areas (**Table 2**), in that order. Considering the sampling effort, urban areas are not only the least diverse but also the least abundant based on number of bats captured, followed by suburban areas and forests (**Table 2**). Even though we found three different assemblages (**Figure 2**) we found no significant correlation between taxonomical α-diversity and the type of landscape (R^2 = 0.04; DF = 1,24; P = 0.24). However, we found species such as *Artibeus jamaicensis*, *A. lituratus*, *Glossophaga soricina* and *Sturnira parvidens* that were recorded in all three landscapes. But there were certain species that were recorded only in certain landscapes, for example, *Glossophaga leachii* in urban areas, *Lonchorhina aurita* in suburban areas, and *Tonatia bakeri* and *Vampyressa thyone* on forests (**Table 3**; **Figure 3**).

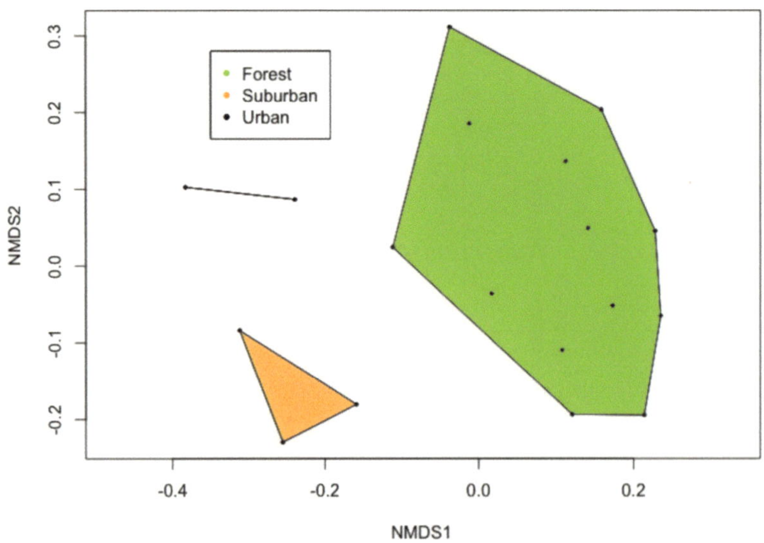

Figure 2.
Non-metric multi-dimensional scaling (NMDS) of three landscapes which represents three different bat assemblages.

	Species	Forest	Suburban	Urban	Total
1	*Artibeus inopinatus*		23		23
2	*Artibeus jamaicensis*	92	48	56	196
3	*Artibeus lituratus*	71	10	62	143
4	*Carollia castanea*	20			20
5	*Carollia perspicillata*	85	23		108
6	*Carollia sowelli*	29			29
7	*Centurio senex*		1	1	2
8	*Chiroderma gorgasi*	2			2
9	*Chiroderma salvini*		3		3
10	*Chiroderma villosum*		9		9
11	*Choeroniscus godmani*		3		3
12	*Chrotopterus auritus*	2			2
13	*Dermanura phaeotis*	8	1		9
14	*Dermanura tolteca*		2		2
15	*Dermanura watsoni*	5			5
16	*Desmodus rotundus*	5	6		11
17	*Diphylla ecaudata*		2		2
18	*Glossophaga commissarisi*	9		2	11
19	*Glossophaga leachii*			1	1
20	*Glossophaga soricina*	3	18	18	39
21	*Hylonycteris underwoodi*	1			1
22	*Lonchorhina aurita*		5		5
23	*Lophostoma brasiliense*	2			2
24	*Micronycteris hirsuta*	2			2
25	*Micronycteris microtis*	4	1		5
26	*Phylloderma stenops*	1			1
27	*Phyllostomus discolor*	4	7		11
28	*Phyllostomus hastatus*	1			1
29	*Platyrrhinus helleri*	6			6
30	*Sturnira parvidens*	1	7	3	11
31	*Tonatia bakeri*	1			1
32	*Uroderma convexum*	18			18
33	*Vampyressa thyone*	4			4
	Total	**376**	**169**	**143**	**688**

Table 3.
Description of the occurrence of the landscapes in which the 33 phyllostomid bats were recorded.

3.2 Activity patterns vs. morphological traits

Considering time (**Figure 4**), we found no significant correlations in any type of landscape: forests and suburban areas (R^2 = 0.00; DF = 1,154; P = 0.23), forests and

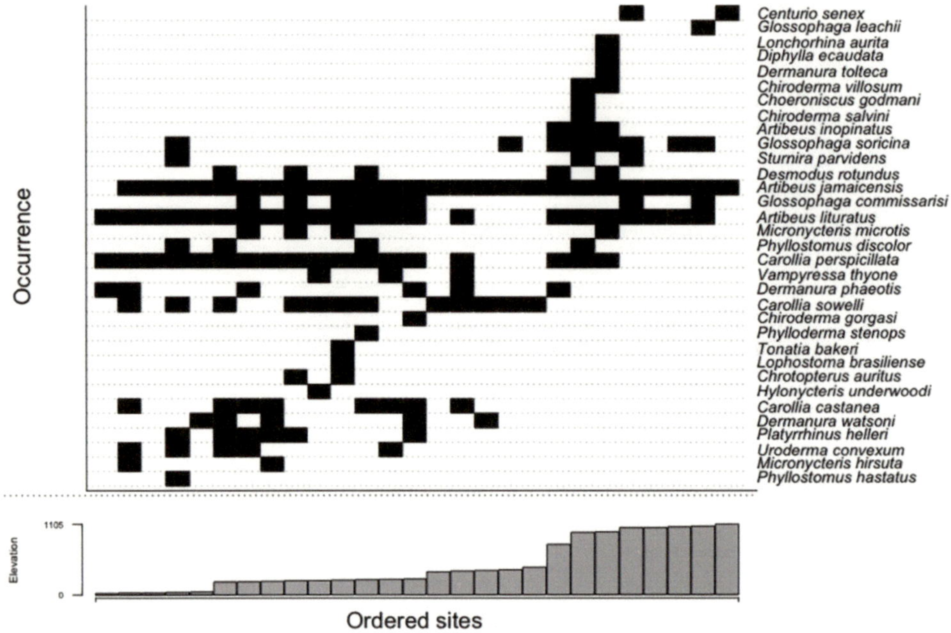

Figure 3.
Occurrence of phyllostomid bats based on the elevation of the studied areas.

urban areas ($R^2 = -0.01$; DF = 1,139; P = 0.62), and urban and suburban areas ($R^2 = -0.00$; DF = 1,139; P = 0.99). In the case of body height means, only the suburban areas and forests have no significant differences ($F_{(2,203)} = 2.3$, p = 0.21; **Figure 5A**), and were divided into two groups urban areas (a) and forests and subruban areas (b). See **Table 4** to see the other p values of this and other analyses. The ear length means were only significant different in suburban and forests landscapes ($F_{(2,165)} = 4.57$, p = 0.05; **Figure 5B**), however, all the landscapes were assigned to the same group (a). When comparing forearm length means with posthoc Tukey tests, landscapes were classified into two groups urban (a) and suburban and forests (a) and only the comparison between suburban areas and forests was not significant ($F_{(2,431)} = 21.41$, p = 0.99; **Figure 5C**). Finally, the moon

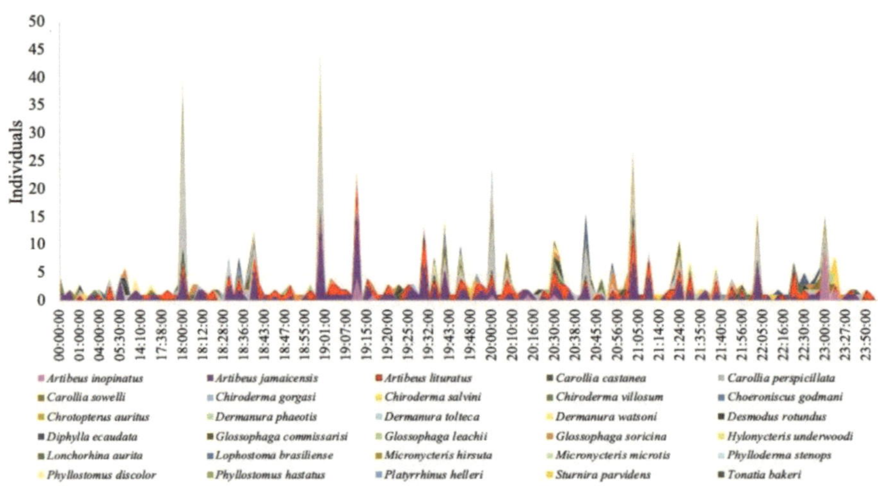

Figure 4.
Time activity patterns of the 33 phyllostomid species recorded.

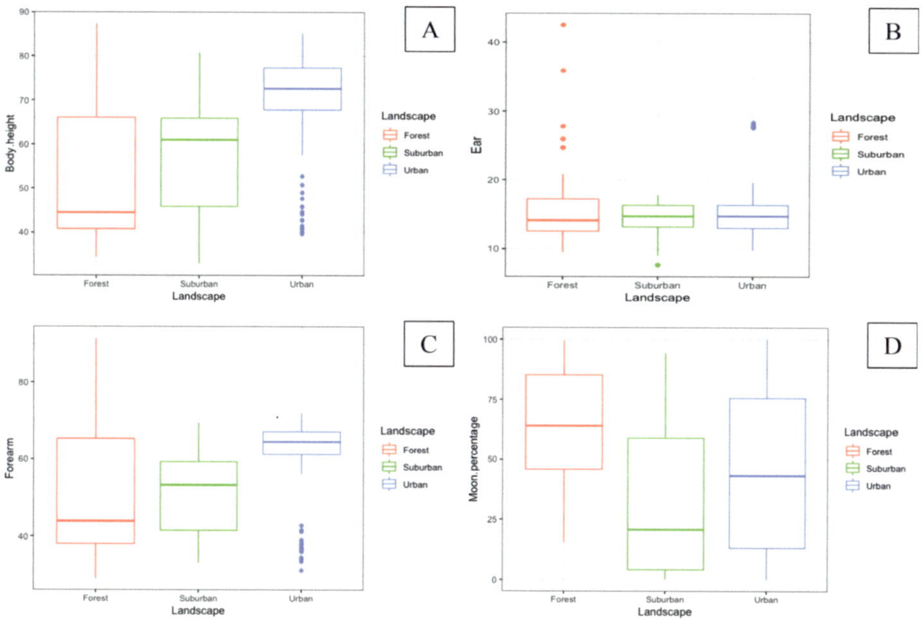

Figure 5.
ANOVA analyses of the body height (A), ear length (B), forearm length (C), and moon percentage (D) time in which phyllostomids were captured.

Landscape	Forests	Suburban	Urban
Forests - FA	—	0.99	<0.01
Suburban - FA	0.99	—	<0.01
Urban - FA	<0.01	<0.01	—
Forests – E	—	0.05	0.17
Suburban - E	0.05	—	0.61
Urban - E	0.17	0.61	—
Forests - BH	—	0.21	<0.01
Suburban - BH	0.21	—	<0.01
Urban - BH	<0.01	<0.01	—
Forests - moon	—	<0.01	<0.01
Suburban – moon	<0.01	—	<0.01
Urban - moon	<0.01	<0.01	—

Table 4.
Statistical results from the comparison of the posthoc Tukey analyses. Abbreviations are as follow: FA = forearm length; E = ear length; BH = body height; moon = moon percentage.

percentage mean in which bats were captured was significant different among all the landscapes (**Figure 5D**), categorized into three different groups a (forest), b (urban areas), and c (suburban areas).

4. Discussion

As expected in Honduras, there is a consistent decrease of phyllostomid bat diversity and abundance from forests to cities. As anticipated, we found that the diversity is less in urban areas (cities) and suburban areas in comparison to forests. However, these remnants of forest are important for bat conservation in urban areas. For example, the high abundance of *Artibeus* and *Glossophaga* in CU-UNAH are, hopefully, helping in the regeneration of trees by seed dispersal and pollination [44] in this small remnant of forest in the capital city of Honduras. On the other hand, suburban areas may also have important species. For instance, Sabanagrande may be the most important area for the conservation of *Artibeus inopinatus* [25, 45], which is considered to be Data Deficient (DD) by the IUCN (International Union for the Conservation of Nature) [46]. That suburb also houses species such as *Micronycteris microtis* which is more typical of forest remnants [47]. In contrast to urban and suburban areas, forests clearly have the highest diversity and abundance of phyllostomid bats. Unfortunately, except in protected areas, these forests are being constantly fragmented and with time, they tend to become agricultural or urban landscapes. Species that cannot acclimatize or adapt to these anthropogenically modified landscapes will disappear. Those that occupy very limited geographic distributions likely will become extinct. Olivier et al. [48] point out that two major drivers of habit degradation are urbanization and agricultural intensification which decrease community stability, including bats.

4.1 How urbanization is affecting bat diversity

Urbanization is the second most detrimental anthropogenic agent of landscape change [49], since bat diversity and species abundance are comparatively lower in cities than in primary forests or rural areas [3]. This is the case not only for Honduras, but worldwide. For example, in Poland, urbanization pressure is a common phenomenon in several protected areas due to the dispersion of buildings and the expansion of summer construction [49]. Additionally, artificial lighting and sound pollution can alter commuting processes in foraging bats, especially sound which has a more deterrent effect for bats than light as some insectivorous bats feed on the insects that are attracted to streetlights [50–52]. Interestingly, habitat degradation affects the diversity of bat communities in more complex ways than simply population stability [48].

Bat response and sensitivity to urbanization varies among species assemblages in urbanized landscapes. In this way species with high tolerance become more abundant and dominant. However, the low diversity and abundance of urban bat fauna can be attributed, at least partially, to a shortage of roosting sites [53]. For a better understanding of bats that do not fly below canopy in urban areas, acoustic monitoring can provide data for species that are rarely captured in mist nets ([17], and see introduction, this chapter). Unfortunately, we have little such data for Honduras as of now.

The fact that forearm length was only significantly different between forests and urban areas and the ear length between suburban areas and forests can be explained from two points of view. First, we found that Sabanagrande has a mixture of the other two assemblages, and even has species that were found only there (e.g., *L. aurita*). Additionally, Sabanagrande has a mixture of ecological characters, and even though it is becoming urbanized, there are still important forest remnants. Similar to Mexico [47], we found that in suburban areas (referred to as intermediate disturbance sites), the predominant species was *Carollia perspicillata*, (subfamily Carollinae). Our results support the hypothesis of Medellín et al. [47] in which phyllostomines (subfamily Phyllostominae) disappear from disturbed areas because of their specific requirements (e.g., foraging activity; [54]) because all the phyllostomines recorded in this study were in forest remnants.

Secondly, the functional traits varied. For example, the well-conserved forests of the RPBR have larger species that were only recorded there (e.g., *T. bakeri*) but in less abundance. In these species their ear length and wingspan are relatively larger, in contrast to those of the large species *Artibeus lituratus*, found in CU-UNAH and Sabanagrande. Ramírez-Mejía et al. [55] demonstrated that the most dominant functional group of bats in their studies were those with intermediate values of body mass and wing morphology, which represents the phyllostomids' adaptive response to landscape degradation.

Species such as *Phyllostomus hastatus*, which is widespread in Honduras, have a larger forearm length and therefore a larger wingspan. Additionally, these species and others of the subfamily Micronycterinae, Lonchorhininae, and Glyphonycterinae, have larger ears because they are gleaners and thus ear length is related to their dietary habits, and these gleaners are more susceptible to the gradual degradation of remnant forests and effects of urbanization (e.g., sound and light pollution) [47]. In other words, phyllostomid gleaners (e.g., phyllostomines) and related subfamilies are indicators of relatively undisturbed rainforests [47, 56]. However, a few species, notably *Micronycteris microtis,* for example, may tolerate moderate levels of urbanization. More ecological data is needed to support this hypothesis.

4.2 From cities to forests: Activity patterns of phyllostomids in Honduras

We hypothesize that New World leaf-nosed bats in forests are more likely to be negatively affected by brightness of the moon because of safety concerns (hunting activities by visually oriented predators like owls) when the moon is brighter [15]. In contrast, urban and suburban areas have equally high light intensity every night (e.g., traffic lights, streetlights, shopping centers, etc.). Another feature that supports our hypothesis is that we found significant difference between all the areas. This is probably because the light intensity of suburban areas is increasing in the same way as in urban areas, and the bats that survive there are able to acclimatize quickly. However, the time patterns were not significantly different among the three landscapes due to the wide range and different foraging behaviors. In general, phyllostomids have an early activity peak and then declining activity through the night [57]. Habitat specialization, nutrient intake, and food procurement are features that are associated with bat success in transformed landscapes [55, 58].

There are two more works describing activity patterns in Honduras. The first one, Medina-Fitoria et al. [59] studied certain areas included in the RPBR, and determined that in the Caribbean slope of Nicaragua and Honduras, mature and intact primary forests are the most important habitats to conserve. They also determined that fragmentation due to extensive cattle farming and agriculture is perhaps the major threat to these forests. And the second study, in the northwestern region of Honduras, in Cusuco National Park, by Medina-Berkum et al. [60] indicated that the presence of *Chrotopterus auritus* and *Trachops cirrhosus* only in the core zone of the park is probably because the core zone still comprises intact and closed forest habitats that may provide a higher abundance of prey for them, and that these species are generally more abundant in undisturbed areas. Primary and pristine forests are clearly among the most important remaining habitats for bat conservation in Honduras, considering that many phyllostomids depend on this type of forest. For example, the core zone and areas nearby of the RPBR are the only ones left in which *Ectophylla alba* may be found in Honduras due to the requirements of the species [61]. (Even in 1967 that was the only habitat in which one of us, RKL, captured *Ectophylla alba*).

Although we predicted that from forests to cities, the diversity of phyllostomids will decrease, this is the first attempt to describe their activity patterns in these areas in Honduras. Considering the extension of forests, Duarte et al. [62] mentioned that 48% of the Honduran territory is covered by forests. With the high rate in which they are being diminished is approximately 23,303.56 hectares per year [63], the probability of losing bat species in Honduras is all too real. On the other hand, there are some species that have been adapting very well, as is the case of *A. jamaicensis* and *A. lituratus*. These are species that were recorded in all three landscapes we studied and were the most common in urban areas. And even if they are considered as tolerant species, they have an important role in urban and suburban area as seed dispersers [64], and probably they should be considered as the phyllostomid species most tolerant to urbanization in Honduras. It appears that species of *Sturnira* and *Glossophaga* are also adapting well to this phenomenon.

5. Conclusion

Undoubtedly, the RPBR is one of the most important regions in Honduras, and probably in Central America, for bat conservation due to the large extensions of pristine forests and the limited occurrence and abundance of certain species (e.g., *Chiroderma gorgasi*) in that region. Unfortunately, even this region will be subject to the effects of encroaching clearing for agriculture and urbanization, if inadequately protected. Indeed, some species become acclimatized to urbanization, and some species, most notably *A. jamaicensis* and *A. lituratus*, now tolerate higher levels of disturbance than many other species. More studies are needed to determine and explain their tolerance to urbanization. There are other species that have very specific requirements, including intact primary forest, for their survival. (e.g., *T. saurophila*). This is the first attempt to study how urbanization is affecting a mammalian group in Honduras and is also the first comparison of bat diversity among three different landscapes. Yet, there are many variables that should be analyzed, compared, and described. For example, we recommend measuring light and sound intensity in urbanized areas and comparing them with those of the forests, to determine more specifically how these characteristics are affecting the diversity and abundance of phyllostomid bats in Honduras. Finally, it is still unknown which morphometrical characteristics are important in explaining the adaption of some phyllostomid bat species to urbanized areas. Another factor is the relative amount of fragmentation in the various areas we studied. There are many gaps in our knowledge of how totally or partially urbanized areas are affecting phyllostomid bats in Honduras, and even though there are some similar activity patterns, we can conclude that their diversity and abundance is decreasing in urbanized areas. Considering the increase of urbanization in Honduras plus the high rate of deforestation (approximately 63.85 hectares per day), a conservation plan for Honduran bats is fundamental.

Acknowledgements

We would like to thank to Alejandro Orellana, Diego Mazier, and Eduardo Ordoñez because most of the field work was done with them and for their comments to this chapter. We are grateful to the staff of El Ocotal, especially to Alejandro Velásquez, for all their support during this research. To the ICF, for the research permit: Resolución–DE–MP–064–2017. We want to thank to all the wildlife and forest manager, native people, civilians, and police and military officer that helped us during all this research. Finally, MATC wants to personally thank Marcia Flores-Casco and her family for their gracious hospitality in Brazil during the COVID-pandemic while he was writing this chapter.

Conflict of interest

The authors declare no potential conflict of interest.

Author details

Manfredo A. Turcios-Casco[1*], Richard K. LaVal[2], Marcio Martínez[3] and Hefer D. Ávila-Palma[4]

1 Departamento de Vida Silvestre, Instituto Nacional de Conservación y Desarrollo Forestal, Áreas Protegidas y Vida Silvestre, Francisco Morazán, Honduras

2 The Bat Jungle, Monteverde, Costa Rica

3 Región Biosfera del Río Plátano, Instituto Nacional de Conservación y Desarrollo Forestal, Áreas Protegidas y Vida Silvestre, Olancho, Honduras

4 Panthera Corporation, Francisco Morazán, Honduras

*Address all correspondence to: manturcios21@gmail.com

References

[1] Jung K, Caragh G. Urbanization and Its Effects on Bats—A Global Meta-Analysis. In: Voigt CC, Kingston T, editors. Bats in the Anthropocene: Conservation of Bats in a Changing World. Springer Publishing Co.: New York. 2015. p. 13–33.

[2] McKinney ML. Urbanization, biodiversity, and conservation. Bioscience. 2002;52(10):883–890. DOI: https://doi.org/10.1641/0006-3568(2002)052[0883:UBAC]2.0.CO;2

[3] Coleman JL, Barclay RM. Influence of urbanization on demography of little brown bats (*Myotis lucifugus*) in the praires of North America. PLoS One. 2011;6(5) DOI: https://doi.org/10.1371/journal.pone.0020483

[4] Dixon MD. Relationship between land cover and insectivorous bat activity in an urban landscape. Urban Ecosystems. 2012;15(3):683–695.

[5] Hale JD, Fairbrass AJ, Matthews TJ, Sadler JP. Habitat composition and connectivity predicts bat presence and activity at foraging sites in a large UK conurbation. PLoS ONE. 2012; 7(3): 7 (3): e33300. DOI: https://doi.org/10.1371/journal.pone.0033300

[6] Pearce H, Walters CL. Do green roofs provide habitat for bats in urban areas? Acta Chiropterologica. 2012;14(2):469–478. DOI: https://doi.org/10.3161/150811012X661774

[7] Macias R. How to see the Austin bats under Congress Bridge – Austin, TX [Internet]. Available from: https://www.tripsavvy.com/austins-bat-bridge-a-viewing-guide-254880 [Accessed: 10-12-2020].

[8] Norberg UM, Rayner JMV. Ecological morphology and flight in bats (Mammalia; Chiroptera): wing adaptations, flight performance, foraging strategy and echolocation. Philosophical Transaction of the Royal Society B Biological Science. 1987 316 (1179):335–427. DOI: https://doi.org/10.1098/rstb.1987.0030

[9] Geggie J, Fenton B. A comparison of foraging by *Eptesicus fuscus* (Chiroptera: Vespertilionidae) in urban and rural environments. Canadian Journal of Zoology. 1985;83:263–265. DOI: https://doi.org/10.1139/z85-040

[10] Kurta A, Teramino JA. Bat community structure in an urban park. Ecography. 1992; 15:257–261. DOI: https://doi.org/10.1111/j.1600-0587.1992.tb00032.x

[11] Lesiñski G, E Fuszara, Kowalski M. Foraging areas and relative density of bats (Chiroptera) in differently human transformed landscapes. Z Säugetierkunde. 2000; 65:129–137.

[12] Gaisler J, Rehak Z, Bartonicka T. Bat casualties by road traffic (Brno-Vienna). Acta Theriologica. 2009;54:147–155.

[13] Novaes RL, Laurindo RS, Dorna RAP, Esbérhard CEL, Bueno, C. On a collision course: the vulnerability of bats to road kills in Brazil. Mastozoologia Neotropical. 2008; 25: 115–128.

[14] Jung K, Kalko EKV. Adaptability and vulnerability of high flying Neotropical aerial insectivorous bats to urbanization. Diversity and Distributions. 2011;17(2):262–274. DOI: https://doi.org/10.1111/j.1472-4642.2010.00738.x

[15] LaVal R, Lawton R. The effect of environmental variables on nightly activity patterns of insectivorous bats monitored over ten years in a tropical premontane forest, Costa Rica. Submitted for publication. Neotropical Biodiversity.

[16] Jung K, Kalko EKV. Where forest meets urbanization: foraging plasticity of aerial insectivorous bats in an anthropogenically altered environment. Journal of Mammalogy. 2010; 91(1): 144–153. DOI: https://doi.org/10.1644/08-MAMM-A-313R.1

[17] Arias-Aguilar A, Chacón-Madrigal E, Rodríguez-Herrera B. El uso de los parques urbanos con vegetación por murciélagos insectívoros en San José, Costa Rica. Mastozoología Neotropical. 2015;22(2):229–237.

[18] Gamba-Rios M. Riqueza de especies, abundancia y dieta de los murciélagos frugívoros de San José. Universidad Latina de Costa Rica; 2006.

[19] Turcios-Casco MA, Ávila-Palma HD, LaVal RK, Stevens RD, Ordoñez-Trejo EJ, Soler-Orellana JA, Ordoñez-Mazier DI. A systematic revision of the bats (Chiroptera) of Honduras: An updated checklist with corroboration of historical specimens and new records. Zoosystematics and Evolution. 2020a;96 (2):411–429. https://doi.org/10.3897/zse.96.51059.

[20] York HA, Rodríguez-Herrera B, LaVal RK, Timm RM. Field key to the bats of Costa Rica and Nicaragua. Journal of Mammalogy. 2019;100(6): 1726–1749. DOI: https://doi.org/10.1093/jmammal/gyz150

[21] Tammaru T, Kulu H, Kask I. Urbanization, Suburbanizarion, and Counterurbanization in Estonia. Eurasian Geography and Economics. 2004;45(3):212–219. DOI: https://doi.org/10.2747/1538-7216.45.3.212

[22] Champion T. Urbanization, Sub-urbanization, Counterurbanization, and Reurbanization. In: Paddison, R. Handbook of Urban Studies. London: Sage Publications; 2001. p. 143–161.

[23] Holdridge, LR. Ecología basada en zonas de vida. Instituto Interamericano de Cooperación para la Agricultura (IICA): San José; 1987; 206 p.

[24] Ferrufino L, Oyuela O, Sandoval G, Beltran F. Flora de la ciudad universitaria, UNAH: un proyecto de ciencia ciudadana realizado por estudiantes universitarios. Revista Ciencia y Tecnología. 2015;17:112–131. DOI: https://doi.org/10.5377/rct.v0i17.2684.

[25] Ávila-Palma HD, Turcios-Casco MA, Velásquez A. The Tome's Sword-nosed Bat (*Lonchorhina aurita*) in Honduras, with new records in Sabanagrande, Francisco Morazán. Bat Research News. 2020;61(1):1–6.

[26] Martinez MA, Turcios-Casco M, Amador S. On the conservation of *Myrmecophaga tridactyla* (Pilosa: Myrmecophagidae) in the core of Río Plátano Biosphere Reserve, Honduras. *Mammalia.* 2020;84(6):581–585. DOI: https://doi.org/10.1515/mammalia-2019-0152

[27] Kunz TH, Kurta A. Capture methods and holding devices. In: Kunz TH, editor. Ecological and behavioral methods for the study of bats. Smithsonian Institution Press: Washington DC. 1988. pp.1–28.

[28] Timm RM, LaVal RK, Rodríguez-HB. Clave de campo para los murciélagos de Costa Rica. Brenesia. 1999;52:1–32.

[29] Medina-Fitoria A. Murciélagos de Nicaragua, guía de campo. 2014. Ministerio del Ambiente y los Recursos Naturales (MARENA): Managua. 279 pp.

[30] Mora JM. Clave para la identificación de las especies de murciélagos de Honduras. Ceiba. 2016; 54(2):93–117. DOI: https://doi.org/10.5377/ceiba.v54i2.3283

[31] Straube FC, Bianconi GV. Sobre a grandeza e a unidade utilizada para

estimar esforço de captura com utilização de redes-de-neblina. Chiroptera Netropical. 8(1–2):150–152.

[32] Sikes RS, The Animal Care and Use Committee of the American Society of Mammalogists. Guidelines of the American Society of Mammalogists for the use of wild mammals in research and education. Journal of Mammalogy. 2016; 97(3):663–688. DOI: https://doi.org/10.1093/jmammal/gyw078

[33] Jost L. Entropy and diversity. Oikos. 2006;113:363–375. DOI: https://doi.org/10.1111/j.2006.0030-1299.14714.x

[34] Moreno CR, Barragán F, Pineda E, Pavón NP. Reanálisis de la diversidad alfa: alternativas para interpretar y comparar información sobre comunidades ecológicas. Revista Mexicana de Biodiversidad. 2011;82: 1249–1261. DOI: https://doi.org/10.22201/ib.20078706e.2011.4.745

[35] Colwell RK. EstimatesS (version 9): statistical estimation of species richness and shared species from sample. 2013.

[36] Colwell RK, Coddington JA. Estimating terrestrial biodiversity through extrapolation. Philosophical Transactions: Biological Sciences. 1994; 345(1311):101–118. DOI: https://doi.org/10.1098/rstb.1994.0091

[37] Amorim F, Rebelo H, Rodrigues, L. Factors influencing bat activity and mortality at a wind farm in a Mediterranean region. Acta Chiropterologica. 2012;14(2). DOI: https://doi.org/10.3161/150811012X661756

[38] Moon Giant. Available at: https://www.moongiant.com/phase/today/

[39] Lichstein JW. Multiple regression on distance matrices: a multivariate spatial analysis tool. Plant Ecology. 2007;188 (2):117–131. DOI: https://doi.org/10.1007/s11258-006-9126-3

[40] Peña-Cuellar ED, Benítez-Malvido J, Avila-Cabadilla LD, Martínez-Ramos M, Estrada A. Structure and diversity of phyllostomid bat assemblages on riparian corridors in a human-dominated tropical landscape. Ecology and Evolution. 2015; 5:903–913. DOI: https://doi.org/10.1002/ece3.1375

[41] R Core Team. R: A language and Environment for statistical computing. Vienna, Austria: R Foundation for Statistical Computing. 2019.

[42] Oksanen J, Blanchet FG, Friendly M, Kindt R, Legendre P, McGlinn D, Minchin PR, O'Hara RB, Simpson GL, Solymos P, Stevens MHH, Szoecs E, Wagner H. Vegan: Community Ecology Package. 2015.

[43] Goslee S, Urban D. The ecodist Package for Dissimilarity-Based Analysis of Ecological Data. Journal of Statistical Software. 2007;22(7): 1–19. DOI: https://doi.org/10.18637/jss.v022.i07

[44] Turcios-Casco MA, Ávila-Palma HD, Ordoñez Trejo EJ, Soler Orellana JA, Ordoñez Mazier DI. Comments on the diet of phyllostomid bats (Chiroptera) in a subtropical dry forest in central Honduras. Studies on Neotropical Fauna and Environment. 2019. 54(3):239–244. DOI: https://doi.org/10.1080/01650521.2019.1663114

[45] Turcios-Casco MA, Ávila-Palma HD, Ordoñez-Trejo EJ, Soler-Orellana JA, Ordoñez-Mazier DI, Meza-Flores DE, Verlásquez A. Rare or misidentifed? On the external identification of the neglected *Artibeus inopinatus* Davis & Carter, 1964 (Chiroptera, Phyllostomidae) in Honduras. Evolutionary Systematics. 2020b;4(1): 35–43. DOI: https://doi.org/10.3897/evolsyst.4.49377

[46] Reid F, Medina A. *Artibeus inopinatus* [Internet]. Available from: https://dx.doi.org/10.2305/IUCN.UK.2016-2.RLTS.T2132A21996586.en [Accessed: 10-12-2020].

[47] Medellín RA, Equihua M, Amin ME. Bat diversity and abundance as indicators of disturbance in Neotropical rainforests. Conservation Biology. 2000; 14: 1666–1675. DOI: 10.1111/j.1523-1739.2000.99068.x

[48] Olivier T, Thébault E, Elias M et al. Urbanization and agricultural intensification destabilize animal communities differently than diversity loss. Nature Communication. 2020;11 (2686). DOI: https://doi.org/10.1038/s41467-020-16240-6

[49] Wilkaniec A, Galecka-Drozda A, Raszeja E, Szczepanska M. Urbanisation processes in Puszcza Zielonka Landscape Park in Poland – and its buffer zone in the context of protection of natural structures. Applied Ecology and Environmental Research. 2020;18 (1):697–712. DOI: http://dx.doi.org/10.15666/aeer/1801_697712

[50] Frank TM, Gabbert WC, Chaves-Campos J, LaVal RK. Impact of artificial lights on foraging of insectivorous bats in a Costa Rican cloud forest. Journal of Topical Ecology. 2019; 35(1);8–17. DOI: https://doi.org/10.1017/S0266467418000408.

[51] Schaub A, Ostwald J, Siemers BM. Foraging bats avoid noise. Journal of Experimental Biology. 2008;211(19): 3174–3180. DOI: https://doi.org/10.1016/j.mambio.2015.02.004

[52] Stone EL, Harris S, Jones G. Impacts of artificial lighting on bats: a review of challenges and solutions. Mammalian Biology. 2015; 80(3):213–219. DOI: https://doi.org/10.1016/j.mambio.2015.02.004

[53] Humphrey SR. Nursery roosts and community diversity of Nearctic bats. Journal of Mammalogy. 1975; 56(2):321–346. DOI: https://doi.org/10.2307/1379364

[54] Bouvet A, Paillet Y, Archaux F, Yillon L, Pascal D, Olivier G, Gosselin F. Effects of forest structure, management and landscape on bird and bat communities. Environmental Conservation. 2016;43(2):148–160. DOI: https://doi.org/10.1017/S0376892915000363

[55] Ramírez-Mejía AF, Urbina-Cardona JN, Sánchez F. Functional diversity of phyllostomid bats in an urban–rural landscape: A scale-dependent analysis. Biotropica. 2020;52:1168–1182. DOI: https://doi.org/10.1111/btp.12816

[56] Castro-Luna A, Sosa VJ, Castillo-Campos G. Quantifying phyllostomid bats at different taxonomic levels as ecological indicators in a disturbed tropical forest. Acta Chiropterologica. 2007;9(1):219–228. DOI: https://doi.org/10.3161/1733-5329(2007)9[219:QPBADT]2.0.CO;2

[57] LaVal RK. Infraspecific Relationships of Bats of the Species. *Myotis austroriparius*. Journal of Mammalogy. 1970;51(3):542–552. DOI: https://doi.org/10.2307/1378393

[58] Welch JN, Beaulieu JM. Predicting extinction risk for data deficient bats. Diversity. 2018;10(63):1–15. DOI: h ttps://doi.org/10.3390/d10030063

[59] Medina-Fitoria A, Ávila-Palma HD, Martínez M, Ordoñez-Mazier DI, Turcios-Casco MA. Los murciélagos (Chiroptera) del Caribe de Honduras y Nicaragua: una comparación sobre su diversidad en cinco diferentes coberturas vegetales, Neotropical Biodiversity. 2020;6(1):147–161, DOI: https://doi.org/10.1080/23766808.2020.1804748

[60] Medina-Berkum P, Vulinec K, Crace D, López Gallego Z, Martin TE.

Community composition of bats in Cusuco National Park, Honduras, a Mesoamerican Cloud Park, including new regional and altitudinal records. Neotropical Naturalist, 2020;3:1–24.

[61] Turcios-Casco MA, Manzanares T, Ávila-Palma HD, Martínez M, Mazier-Ordoñez DI. Reproductive, morphometric, and roosting description of the Honduran white bat, *Ectophylla alba* (Chiroptera: Phyllostomidae), in Honduras. Mastozoología Neotropical. 2020c;27(1):172–176. https://doi.org/10.31687/saremMN.20.27.1.0.07

[62] Duarte EA, Orellana Díaz O, Maradiaga IE, Casco FL, Fuentes DE, Galo AJ, Avilés PE, Araneda FM. Mapa forestal y de cobertura de la tierra de Honduras: análisis de cifras nacionales. 2014. Instituto Nacional de Conservación y Desarrollo Forestal, Áreas Protegidas y Vida Silvestre (ICF)/ Programa Regional REDD/CCAD-GIZ, Honduras. 24 pp.

[63] ICF (Instituto Nacional de Conservación y Desarrollo Forestal, Áreas Protegidas y Vida SIlvestre). Anuario estadístico forestal de Honduras. 2018. ICF: Tegucigalpa; 142 pp.

[64] Lobova TA, Geiselman CK, Mori SA. Seed dispersal by bats in the Neotropics. 2009. New York (USA): The New York Botanical Garden Press. 471 pp.

Chapter 9

Meaning and Health Impact of Food: Historical and Ecological Analysis

Rebeca Monroy-Torres, Graciela Velázquez-Delgado, Erika Carcaño-Valencia and Gilber Vela-Gutiérrez

Abstract

Currently the world is facing a viral contingency that has exposed the vulnerability in which the human being is in the face of the alarming statistics of obesity at all ages, increasing the numbers of diabetes, cardiovascular diseases, and cancer mainly. But not only health has suffered a deterioration worldwide but also the environment with impacts on the availability and quality of water, air pollution and soil deterioration. In México, the food culture has undergone changes derived from greater accessibility to industrialized foods, less physical activity, stress, replacement of consumption and the use of local foods. Currently in Mexico actions are being implemented to rescue agroecology, gastronomy, and food culture. Therefore, this chapter will have the purpose of integrating a historical, ecological and health impacts analysis with what food means and its source or origin for people.

Keywords: sustainable diet, food culture, obesity, industrialized food

1. Introduction

The economic, political, social, environmental and health impacts of the SARS-COV2 virus pandemic as well as the disease generated (COVID-19), worldwide are and will continue to increase, affecting the most vulnerable population. Before the contingency, there was already a deterioration in the health of the population with an increased prevalence for chronic degenerative diseases associated with obesity and unhealthy lifestyles (sedentary lifestyle and consumption of high-energy foods), as well as other problems of health (short stature, malnutrition, anemia, mainly). We are in a century with the greatest scientific advance, but with a population that suffers from hunger and at the same time with the highest figures in the loss and waste of food in the world [1].

In Mexico, according to the recent National Health and Nutrition Survey (ENSANUT-2018 for its acronym in Spanish) [2], there was an increase in the numbers of diabetes (10.3%), arterial hypertension (18.4%), dyslipidemias (19.5%) as well as one of the main risk factors: obesity and being overweight. Obesity and overweight have been exacerbated mainly in adolescents and young adults aged 20 years and over, where for the latter group, for every 10 adults 8 are overweight or obese. Another problem is food insecurity in households in Mexico, which was

70% for 2012 and for 2018 44.5%. Food insecurity had a decrease of 14.5%, but this increase in households with food security cannot be attributed to an improvement in access to nutritious food, since for this same year the consumption of industrialized foods with high energy intake, fat saturated and added sugars increased as well. This deterioration in the health of the population is one of the reasons why the worldwide morbidity and mortality figures for COVID-19 were high for the population living in extreme poverty and with the presence of the diseases [3, 4]. The health impacts have occurred as well as the deterioration of the environment, they have been increasing and there are several theories, including an obsolete economic model where vital resources such as water have been depleted. For Mexico, the lack of access to water in quantity and quality has been the result of mismanagement of the resource, mainly with overexploitation by the industrial sector and the lack of a reform to the National Water Law [5].

The models of production and of labor and economic organization have led to changes in the food culture and, therefore, to people's health and the deterioration of the environment. A production model that has generated greater contamination with impacts on the soil, generating more deteriorated crops, loss of health, which prevents ensuring the sustainability of future generations [6]. Several have been the industrial sectors whose production processes have generated great impacts on health and the environment. The figures for obesity and diabetes in Mexico and in the world have increased in the last three decades due to the construction of obesogenic environments [7]. The role of food has been part of this process, with access to more processed foods and more agrochemicals.

The soft drink industry has been one of the industries with the most evidence of its impacts not only on health but on the environment, mainly in water. For example, for every liter of soft drink produced, 1.7 liters of water are required [8]. In the 70's, Mexico was already the second place in the consumption of soft drinks and scientific evidence has shown that its consumption (sugary drinks) is a risk factor for developing obesity and diabetes by contributing 9.8% of the energy intake in the Mexican diet, while 16% are desserts, cookies, sweets [8–10]. Other data that have facilitated their access has been the price of the soft drink and the size of the portion that is sold in Mexico. The price is also different in rural and urban areas. For example, in 2017, for every 2.5 L of soft drink, the cost was 0.80 USD or 0.66 €, while in urban areas it was 1.30 USD or 1.06 € (**Figure 1**). The same happens with the 600 mL presentation, the cost was higher in urban populations,

Figure 1.
Soft drink cost in rural vs. urban areas in México. Source: Courtesy Monroy-Torres R (2017). http://bombochis.blogspot.com/2017/.

	Mexico	Germany
Maximum amount of soft drink sold.	3 L (0.80 USD or 0.66 €)	2 L (2.14 USD or 1.74€)
Cost per 600 mL [a,b,c]	0.75 USD (0.62 €).	2.43 to 3.65 USD (2–3 €)
Liters per capita consumed (Ref. [11], d).	142 to 233	97 to 135.5
Water required to produce one liter of soft drink [11]	1.71 L	
Global position in bottled water consumption [11]	2	6
Diabetes in the adult population (%)	10.3%	7.4%
Obesity and overweight in adults 20 years and older [2]	75.2% 2018	82.4% 2016

Source: Websites: a) https://super.walmart.com.mx/refrescos/refresco-coca-cola-botella-de-600-ml/00000007500761 b) https://marginalrevolution.com/marginalrevolution/2009/09/why-is-cocacola-so-expensive-in-europe.html; c) https://www.expatistan.com/price/coke/berlin; d) Liters per capita according Euromonitor International 2013. Dollar and Euro Exchange, on december 28, 2020.

Table 1.
Soft drink consumption and health statistics in Mexico and Germany.

which may explain its accessibility in all regions of the country. In **Table 1** a comparative summary of some data on soft drinks in Mexico (Latin American Country) is presented. and contrasting it with a European country like Germany.

1.1 A historical perspective

Today the diet is different from what was used in other times. Not only have the inputs with which food is prepared have changed, but also the eating patterns, the way they are prepared and the value of these. All this has had an impact on the nutrition of the population. This change has been caused by different factors that have to do with cultural miscegenation and with the cultural influences that have been established in Mexico over time, even with the scourge of natural phenomena such as droughts, floods, and diseases.

Environmental devastations have been constant throughout history; one of them occurred precisely in the 16th century, when the Spanish arrived. This does not mean that the environment had not changed previously, but rather that, according to specialists in environmental history, they were of lesser magnitude than those that occurred after the arrival and settlement of the Spanish. They brought new ways of producing food by introducing both sowing and cattle grazing [12].

Regarding the diet consumed by the natives, Guillermo Santamaría, an Augustinian friar from the 16th century, affirmed that the Chichimeca Indians consumed wild fruits and roots, in addition to hunting products and sometimes fishing. What they most consumed was corn, beans, squash, chili, tunas, nopales, pitahayas, xoconoxtle, sweet potatoes and mesquite pods, from which they even made bread that they could keep for a long time. The diet was based on plants that were collected by women, but also on those that were planted in cornfields or family gardens. In addition to the above, they also fed on game animals such as snakes, lizards, rabbits, hares, and deer that men hunted in the field. Pulque or tuna or mesquite ferments were used as drinks. Livestock was an important source of food, but also of devastation of the environment, since they ate both the plants of the cornfields and those that grew wild, destroying tunnels, nopaleras, mesquitales and magueyales causing a shortage of food [13].

As can be seen, little by little the diet was modified over time, since, from being made up of legumes, fruits, vegetables, and cacti, it was transformed, and various products of animal origin were added. Among the incorporations of great importance were the sugars with which jams, crystallized sweets, eggnog, and desserts are prepared. Flours, with which a great variety of breads, cakes and cakes were prepared to which seasonal fruits were added. Another incorporated ingredient was lard, which accompanied food not only as an element for cooking, but as an ingredient that added flavor. This was included abundantly in almost most of the cooked foods. He even added to foods that already contained fat, increasing their fat content, mainly those of the saturated type. This had a reason, since fat was an ingredient valued as something that provided energy to individuals, so it is recommended that as much of it be eaten and even the sick was recommended to eat pots or lamb broths to soon to be reestablished [14]. Currently we know that these eating behaviors have a risk to people's health and have probably been the cause of several diseases, but at that time they were not considered that way.

Wondering European travelers constantly related the quantity and variety of fruits and vegetables that were observed throughout New Spain. This shows the fertility of the soil and there was an intensive agriculture based on the rain cycle in each region. But all this was modified by climate change, since the rainy seasons have been significantly altered and that has led to the implementation of irrigation systems that have ended the water reserves in the subsoil. The climates affected food production, both the torrential rains that flooded the fields and the droughts that regularly hit some parts of New Spain. All these natural disasters caused food shortages and the few that did exist were priced at high prices. Because of the scarcity, the population modified their diet, and perhaps these circumstances probably caused the population to adopt a diet based on plants, vegetables and cacti [14].

During New Spain there was a great biodiversity of food and with them its greater accessibility but that depended on the regions and climates, the most common was that people ate prepared food per day. But if there were perishable or more decomposing foods, then a preservation process was applied to them to be able to consume them later.

Food depended for the most part on what was grown and harvested locally, as there were several factors that made it difficult to transport food for mass consumption, either because the food was not kept for so long or because the roads were not conducive to transport them agilely. Most of what was produced was concentrated in the cities, and not only the inhabitants of the city but also those of the rancherias and nearby towns came to it to trade food. In the Bajio area and in others of New Spain (today Mexico) such as Puebla, Michoacan, and Jalisco, they were mainly dedicated to the cultivation of cereals such as wheat and corn, and this determined the diet of the population, since they ate what they harvested, and what was left they sold to buy the rest that they needed. All these limitations made the diets very unbalanced, since most of the population did not have access to all kinds of food, and those who had a comfortable economic position ate the diet that they valued as adequate: fats, sugars and flours in excess causing some diseases [14].

Because the population had an inadequate diet, it was more vulnerable to being attacked by epidemics. For example, cholera was a feared disease for its symptoms (diarrhea and vomiting) that generated high mortality in the population, which is why some foods were prohibited. In epidemics it became common for diets to be changed to control the disease; for example, for the cholera epidemic that affected the state of Guanajuato in 1850, it was recommended that foods laden with spices and fats should not be eaten, that the meats were not salted, salted, or highly seasoned, and that pork meat should not be eaten, the lard bread, the vegetables, the fruits, the seeds that have skin or skins and the milk. Among the foods that were allowed were tender meats, such as chicken, hen, mutton, tender veal, roasted, cooked, and simply seasoned: well-boiled bread, rice, or noodle soups, atole, sago, champurrado and chocolate [15]. All the above tells us about the valuation about healthy or adequate to eating.

Later, with the French influence that the Porfirian regime promoted in Mexico, other culinary practices were introduced. French rolls or brioches, desserts and wines were incorporated. However, this French-style food practically had its impact on the upper social classes. In this national historical period, the commercialization of food and its derivatives was promoted; this increase was due to the construction of the railway in which large quantities of products can be transported in a shorter time [16].

Already in the twentieth century in Mexico, a boom began with the advent of industrialized products. At the beginning of the century, the influence of American food was more in the northern states, but little by little its commercialization spread throughout the country. These products included soft drink, which was gradually incorporated as part of the diet, as well as foods that contained high energy density foods [11, 17].

With technological development, refrigerators were manufactured in the second half of the 20th century, which generated a change in the way of preserving food, replacing the drying, salting, smoking and fermentation of food [18]. And although these techniques were not stopped, the truth is that the kitchen was revolutionized because food can be preserved in greater quantity and for longer. However, this brought both positive and negative changes in the way of cooking and eating, because with this they stopped consuming fresh foods such as fruits and vegetables and meats that were acquired daily. Large quantities of industrialized bread canned, and frozen products also began to be consumed. The milk is industrialized, and the consumption of natural milk almost completely disappears. Animal fats were changed to vegetable fats. But one of the most significant changes will be the use of corn flour for the rapid and industrialized preparation of tortillas, leaving nixtamal almost in disuse. The industrialization of income a homogenization of food throughout the country, reducing over the years the consumption and use of regional and local foods.

Subsequently, with the intense commercial exchange between Canada, the United States and Mexico through NAFTA or NAFTA (for its acronym in English) there have been accelerated changes in what is consumed, eaten and cooked. All this

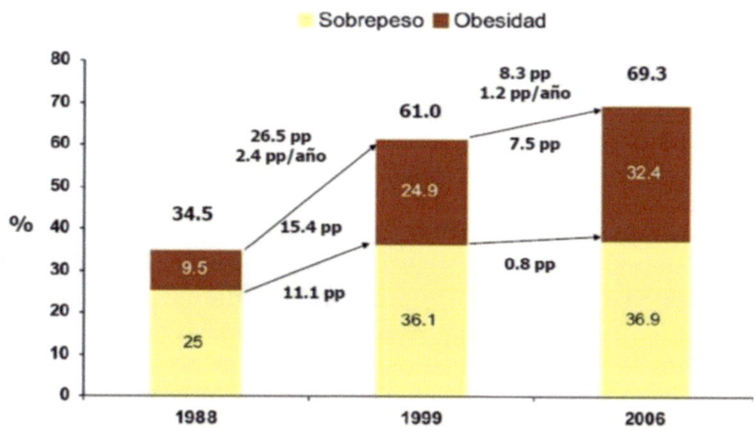

Figure 2.
Increase of overweight and obesity in Mexico from 1980 to 2006. Source: National Agreement for food health. Strategy for overweight and obesity 2010 [20].

has resulted in culinary practices being transformed. Industrialized products are more at hand to be consumed. **Figure 1** presents a chronology of health statistics 10 years before and after the entry into force of NAFTA (1994) [19].

It should be noted that the change in diet that has originated is not exclusive to Mexico but to a global scope, in which products, foods and culinary tastes have been standardized. As we previously pointed out, this type of industrialized diet based on sugars, flours and fats has resulted in high levels of obesity, diabetes, and hypertension, as we will see in a moment. In 1988, overweight and obesity was 34.5% and 5 years after signing the FTA (1999), overweight and obesity increased to 61% and whose trend has been to increase to 75.2% of overweight and obesity combined. The population in Mexico as represented in **Figure 2** [19, 21].

As of the Free Trade Agreement, there was an increase in the price of the basic food basket (fruit, vegetables, beans, corn, beef, chicken, and eggs) while there was a decrease in the price of industrialized food [17].

2. Indicators (economic, environmental, food and health)

The impacts generated on the environment and health attributed to human action have been widely documented. There is data that shows that a total of natural resources exceeding 60 billion tons per year have been extracted so far in the 21st century, at the same time the richest 10% of the world's population monopolized 40% of the energy and 27% of the materials [22].

Climate change has been another factor that adds to the environmental challenges in the era of the so-called "anthropocene". In recent decades, climate changes have shown their effects on ecosystems, productive sectors and society. In agriculture, these impacts are reflected in the yield of crops, affecting in a more profound way the sectors of the population living in poverty, since they have fewer possibilities of generating resilience strategies in the face of predatory rules imposed by the market. In the same way, the possibility of protecting the places they inhabit due to hydrometeorological phenomena is annulled.

Diseases related to the effects of climate change such as the case of vectors such as the mosquito that transmits dengue and malaria, likewise, increases in mortality associated with extreme heat events have been documented [23]. The increasing and accelerated pace of demand for natural resources and energy has generated negative impacts on ecosystems, as well as serious socio-environmental consequences. As biophysical limits continue to be transgressed, we will see more and more irreversible and damaging effects on health, nature and coexistence systems, with the danger of falling into a deterioration that generates new pathologies, ailments and impossible health problems to elucidate in the short and medium term. Hence the need to pay special attention to this process to carry out actions that allow us to have a more harmonious relationship with our environment and therefore a better quality of life.

The latest estimates indicate that in Mexico about 50% of natural ecosystems have been lost, the main effects have been in ecosystems that are more productive, accessible and with better soils. The forests have been the terrestrial ecosystems of the country that have suffered the greatest disturbances due to human activities, both in eliminated and degraded areas [23]. The factors associated with the loss of habitat are related to the change of land use to give way to agricultural, livestock, industrial, tourism, oil and mining activities, among others [24].

In addition to the above, the water quality indicators show that 12 million people lack access to drinking water, 102 of the 653 aquifers in Mexico are overexploited, 46% of the water is lost due to leaks in the supply networks, 80% of water bodies present some type of contamination by industrial discharges [25]. Water contamination is a public health problem, digestive system diseases derived from consuming contaminated water are the third most important cause of infant death in Mexico [26, 27].

2.1 Environmental and health situation in Mexico

Likewise, the generation of urban solid waste (MSW) in Mexico reached 53.1 million tons in 2015. The increase in the generation of urban solid waste can be explained as a result of multiple factors, recognizing urban growth among the most important, industrial development and the change in consumption patterns of the population. The environmental and health consequences of the increase in MSW are presented through factors such as: the generation of biogases, contamination of soils and bodies of water, as well as the proliferation of harmful fauna and transmission of diseases [26, 28].

The global burden of diseases attributed to environmental factors is estimated at around 25% of the total for the general population and around 38% for the child population. The increase in non-communicable diseases and obesity rates in Mexico are alarming, there are several factors that have triggered this phenomenon, one of them is the high consumption of sugary drinks and ultra-processed products with high amounts of fats or sugars, low in content. Fiber and refined flours [29].

In Mexico, in 2016, 72.5% of adults were overweight and obese. Obesity increases the risk of suffering from other diseases such as diabetes mellitus, ischemic heart disease, hypertension, cerebrovascular diseases, cancer, among others, which decrease the quality of life and increase the risk of death among those who suffer from them. Obesity also represents high medical costs, estimated at 151,894 million pesos in 2014 alone, which is equivalent to 34% of public spending on health and causes an estimated loss of productivity of 71,669 million pesos per year [2, 29].

Another problem related to health and the environment is exposure to air pollutants in cities and exposure to chemicals such as pesticides. It is estimated that 42% of chronic lung diseases are due to environmental risk factors related to occupational exposure to dust and chemicals, as well as air pollution in closed spaces due to the burning of solid fuels such as the use of firewood for cooking or as heating [23].

Regarding pesticides, the most frequent use of these occurs in agricultural activity. The prevailing agro-industrial system leads to the intensification of food production and this resulted in the excessive use of pesticides. In Mexico, pesticides are used that are prohibited in other countries, without rigorous controls being carried out for their use, supervision or information that account for the risks and level of danger that they bring with them, in such a way that we are directly or indirectly exposed to the harmful effects that they generate [30].

The soil is the natural resource from which food comes and, the conservation of the earth is thanks to the biogeochemical cycles that allow the movement of elements such as nitrogen, carbon, oxygen, sulfur, water among other elements in the form of cycles that it occurs through living beings and the environment. These cycles have been altered by atmospheric pollution, water exploitation, over-paving of the soil, the intensive use of agrochemicals derived from an economic system and lifestyle of people that has not allowed biogeochemical cycles to be generated. For example, it is known that the long-term use of agrochemicals as fertilizers has impacts on the diversity and density of soil bacteria. The justification for its use derives from achieving greater production to feed a growing world population, but this is only achieving a loss of crops due to a deterioration of the soil and its biogeochemical cycles [31, 32].

The problem shows us that the idea of unlimited economic growth, production dynamics, as well as consumption patterns are increasingly unsustainable; if the trend described above continues, it could put life at risk in all its aspects. Dimensions. We are faced with a scenario that urges us to think of an economy that has life support at its center. Mexico occupies the 7th place as a world producer of vehicles and with an important participation in the market, as can be seen in **Table 2**.

Figure 3 shows the 2010 water map, where the problems of water availability in the country can be observed, and in regions where economic productivity is higher, so it is important to consider the impact of the automotive industry, of soft drinks, sugar, mining, paper, textiles, among the main ones. These industries affect the availability of water in addition to generating many pollutants to the water [11, 33]. On the other hand, overexploitation of water leads to the presence of pollutants such as arsenic and fluoride, where there are several regions in Mexico where high levels of arsenic and dental fluorosis have been found impacting the health of the population [26, 27].

7th world producer of vehicles in general with 3.6 million vehicles [4,5].	**Of the main assembly companies, 21 have a presence in 14 states of the country [9]**
7th world producer of light vehicles with 3.4 million vehicles [4,6]	More than 300 first level suppliers (TIER 1) of the terminal industry [5]
4th Exporter of light vehicles With 2.8 million vehicles [7]	Creation of 81,927 jobs in the automotive terminal industry [10]
5th world producer of heavy vehicles with 191,000 vehicles [4,5]	Participation of the automotive and auto parts industry: National GDP[10]: 3% Manufacturing GDP[10]: 18% Foreign Direct Investment[8] (FDI): 20% Total, exports[9]: 27%
4th Exporter of heavy vehicles with 156,900 vehicles [5]	

Source: Adapted and extracted from ProMéxico with information from [1]MarketLine, estimates [2]OICA [3]Fortune Global 500, 2015 [4]OICA [5]ANPACT [6]AMIA [7]Global Trade Atlas y AMIA [8]Ministry of Economy [9]ProMexico [10] National Institute of Statistic and Geography (INEGI-by its acronym in Spanish).

Table 2.
Data on vehicle production in Mexico.

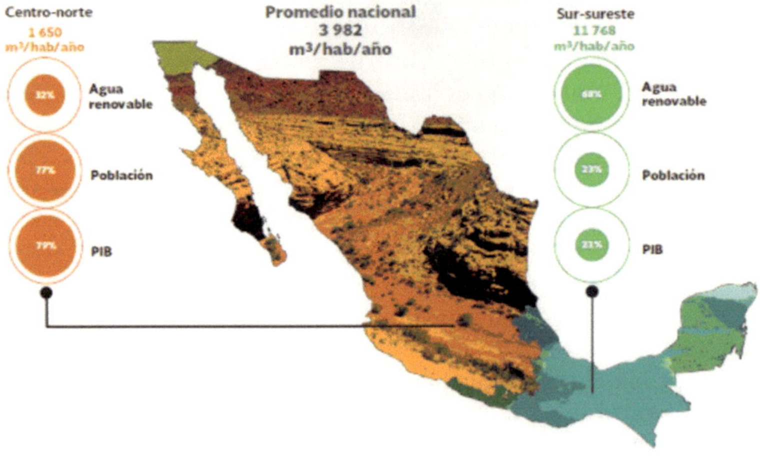

Figure 3.
Source: CONAGUA. www.dof.gob.mx/nota_detalle_popup.php?codigo=5339732.

3. Technology and its ecological, food and health impacts: analysis and proposals

We understand by technology, in a general way, the application of knowledge to generate new methods, processes, services and devices. Technological innovation, on the other hand, is the transformation of an idea into a product, equipment, or operational process, including new forms of social organization. When studying technological innovation, the analysis of the generation /adoption /implementation/monitoring and evaluation processes of technologies should be included [34].

3.1 Impact of technology on ecology

The extinction of species in the Anthropocene, global climate change and the damage to natural ecosystems caused by human activities, are part of an environmental and socio-ecological crisis that manifests itself at local, regional, and global scales. This crisis calls into question the social coexistence and the bases of the very existence of humanity. The Sustainable Development Goals (SDGs) are also indicative of a general framework of poverty, hunger, food insecurity, malnutrition, unsustainable agricultural practices, inequity (economic, social and gender), waste of natural resources, excessive energy consumption and violence and insecurity [34].

3.2 Impact of technology on food production

Crop production technology has changed significantly in recent decades. First, between 1980 and 1990 the direct sowing technique expanded rapidly to replace conventional tillage; later, production was intensified through a greater use of agrochemicals, mainly fertilizers; later transgenic crops were incorporated; and, more recently, differential management by environments, also called "precision agriculture" [26], began to spread.

Expected changes in income and demographics will lead to increased consumption of meat, dairy products, fruits, vegetables, and edible oils, increasing demand for agricultural raw materials. More consumers will enjoy an economic and lifestyle situation that will allow them to buy more processed and packaged foods, as well as a growing variety of convenience and luxury food items, but which will not necessarily increase the demand for agricultural raw materials. The ability of the agricultural and food industries to continue to respond to the undoubted increase in demand over the coming decades will largely depend on the increased application of existing technologies, as well as the exploitation of new and innovative technologies [27].

Since its inception, industrial development has polluted the air, water, and land, irreversibly affecting, in some cases, ecosystems. The rapid and excessive use of natural resources and the disregard for preventive measures have favored environmental pollution. In Latin American and Caribbean countries, the overlap of old communicable diseases with new chronic degenerative diseases is notable, together with environmental risk factors or lifestyles that are the cause of increasing morbidity and mortality, and the increase in the costs of health care and decreased productivity and quality of life [28].

The gap is growing, in the generation and application of technology between countries of the center and the countries of the periphery substantially increased the magnitude of poverty in developing countries. Thus, despite spectacular increases in agricultural productivity in recent decades, undernutrition persists in many nations in Asia, Africa and to a lesser extent in Latin America.

Undoubtedly, food insecurity, more than a production problem, is a problem of access to available food [29].

According to Lorenzana [29], world food production grew at an unprecedented rate because of the application of the modern system between 1950 and 1970. By then, energy was cheap and there was worldwide the possibility of expanding areas cultivated. This expansion in food production occurred primarily in industrialized countries, especially the U.S.A., Canada, Australia, and New Zealand. The high subsidies that producers enjoyed in countries like the U.S.A. and Canada made it possible to offer abundant and low-priced food, not only for domestic consumption but also for export. Developing countries, with levels of technological advance far below industrialized countries, took advantage of the low international prices of cereals and later of oilseeds. It was simply cheaper to import the staples than to grow them. The fertile lands in the Third World countries were dedicated to the sowing of non-basic products for export, generating foreign exchange that industrialization required. This is how, during those two decades, there was food in abundance and at low prices.

Several reports indicate that the modernization of food production has had negative consequences on food security. Long ago the human diet was varied, it was made up of various species of plants and animal species. The development of technology and agricultural production methods generated a tendency to focus on the most productive and profitable species, being those that were commercialized in urban areas more profitable, with the rural areas having greater purchasing power. Changes in lifestyles, mainly in large cities, coupled with economic fluctuation, contributed to a more monotonous diet. With technology transfer in developing countries, traditional consumption patterns increasingly resemble Western patterns [29].

For some years and perhaps due to the general recognition of the role of diet in the achievement and maintenance of health, an intense search began, in most cases with great scientific rigor, on food and its effect on health [18].

3.3 Technological and health trends

Some trends in which not only the search for healthy food is combined, but also the possibility of eating properly in today's difficult world, show that the general public looks for less processed foods with a similar appearance and quality to freshly prepared ones. These include fresh or minimally processed foods, prepared or precooked dishes (refrigerated, frozen), semi-prepared or precooked products that only require heating for consumption and "fast food" in which it is valued that it is quick to consume, easy to carry and which are also healthy products [30].

Technologies have been developed focused on the maintenance or preservation of food, whose objective is the search for alternative heat treatments and the development of non-thermal preservation treatments, to achieve healthier products, with a longer shelf life, and to instead offer the consumer food with minimal processing. These treatments include electrical pulses that are based on the exposure of a food to an electric field, achieving that of microorganisms by destroying the cell membrane, high pressures in which the high hydrostatic pressure has partial sterilization effects, obtaining products of optimal microbiological quality. With few modifications in aroma, flavor, and nutritional value [30].

Other technologies used in food preservation are irradiation, ideal for solid or even frozen foods, the pulses of light which, as the name implies, are flashes of light of great intensity and short duration that eliminate microorganisms and bio conservation in which the Normal bacterial flora of food is controlled to increase its shelf life. It can also favor the growth of a natural microorganism, to limit the growth of others [30].

3.4 Loss and waste foods and its reduction

Recent estimates indicate that a third (30 to 40%) of the food produced globally is lost and wasted in the global food system. Food insecurity is an issue that motivates greater production and quality of food through sustainable ways, but at the same time allows reducing food loss and waste [31, 32].

Along with reducing food loss and waste (FLW) are new ways to preserve food for more. For example, the dehydration of fruits and vegetables with CO_2 that excludes negative aspects of conventional dehydration, resulting in a product with better quality, low refrigeration cost, the nutritional quality is maintained and when the product is rehydrated it acquires the appearance of a fresh product [31].

The reduction of food loss must be sought from initial production (farm) to final consumption at home, through short marketing chains. One option is to obtain foods from local markets and thereby reactivate local economies [31].

On the other hand, public policies should be established that motivate both the government and the private sector to develop infrastructure for roads, transportation, storage, and refrigeration facilities, which allows reducing food losses. In addition, sensitize the population to avoid the compulsive purchase of food and provide information and knowledge that allows the actors of the agri-food chains to have standards of safety and hygiene, guaranteeing quality food. However, it is necessary to reactivate the exchange of food, at the local level, to reduce the loss of foods. Likewise, carry out research to develop innovations that allow the elaboration of food products from the remnants [31, 32].

The challenge is not only the production of food for a population in constant growth, but also how to ensure sufficient clean water, agricultural land, energy, and labor, in such a way that the adverse effects on the environment are reduced and satisfying the basic needs of present and future generations [31]. Despite the technological development generated in recent decades, there is still a deficiency in food production, as well as in its transformation; on the other hand, the immense amounts of FLW in the entire food chain causing problems of food insecurity, in addition to serious contamination problems, which opens a huge possibility for the development of strategies to improve the food security conditions of the population.

4. Health impacts: an analysis of the main determinants

The changes in health with an increase in the prevalence of chronic degenerative diseases are increasing according to the statistics at the world and national level. These results are a summary of the main problems that continue to be sustained in the country such as obesity, overweight, dyslipidemia, hypertension and an innovation in this survey is the data on blood lead levels and the frontal labeling questionnaire (**Table 3**). For physical activity, only 29% of the population performs physical activity for less than 150 minutes per week and the other extreme, 28.1% performs 1680 minutes per week. The survey separates figures for food security and breastfeeding by urban and rural area, but those presented in the table are for urban areas.

For the consumption of fruit, vegetables, and legumes, they were below 50% of the different age groups except for the consumption of legumes that was 54.4 for the group of 20 and over. For non-recommended foods, non-dairy drinks sweetened or with added sugars were among the highest for the population of all age groups without difference, being in a range 83.3% for the group of 1 to 4 years to 85.5% for the group 20 and over [33].

On health issues and at the time this review is being carried out, we can observe that sufficient evidence has been generated on the impacts on the industrialization

Indicators	ENSANUT* 2012 Percentage (%)	ENSANUT* 2018 Percentage (%)
Food security	30	44.5
Diarrheal disease from 0 to 4 years old.	11	11.8
Diabetes in the population aged 20 years and over	9.2	10.3
Hypertension in the population aged 20 years and over.	16.6	18.4
Cholesterol and triglycerides in the population aged 20 years and over	13	19.5
Tobacco habit in the population 10 to 19 years old	1.5 a 9.5	5.1
Exclusive breastfeeding	14.4	28.6

*Source: Monroy-Torres R [33]. *National Health and Nutrition Survey [ENSANUT].*

Table 3.
Comparison between main results of the ENSANUT 2012 vs. 2018.*

of food and accessibility to them by the population, a recent study found that the consumption of two glasses a day of soda was associated with higher mortality [OR, 1.17; 95% CI: 1.11–1.22; P < 0.001], mainly in participants who presented obesity, although not for those who were overweight [34].

4.1 The meaning of eating

What are the impacts of the industrial revolution up to the era of globalization? Where mobility and commercial exchange (products, services, and food) is more efficient while work and family life is more sedentary and inheritance or eating habits are acquired from social networks, television, different media [17]. The integration of sugar and with it a boom in sugar mills until its replacement in the era of biofuels (high fructose corn syrup). Although the factors associated with obesity and chronic diseases are multiple, the medical literature shows that the consumption of sugary drinks and diets with high energy density are two of the risk factors that contribute to the excess burden of these diseases.

The act of eating is complex, food has a vital component by providing nutrients for the functioning of the human body or organism; but food has cultural meanings, attributes, it is something more than the sum of food, something more than its nutrients or culinary preparation or knowing the origin of these [17]. This meaning will depend on internal and external factors, the latter such as social, political and economic factors that give meaning and that act becomes beyond thinking only about nutrients and health. This must be considered to integrate the meanings.

5. Conclusion

Talking about the meaning of food and food is talking about the main diseases that occur in this XXI century as well as before, during and after the contingency by COVID-19, such as the impacts that an obsolete economic system has had on reduce people's health and generate an impact on the environment. The multidisciplinary approach allowed to have indicators with which the origin and trajectory of the health-disease processes must be understood from the meaning of food, such as: an economic system and production of goods and services different from that of 40 years ago. That has changed the lifestyle of people, with greater sedentary lifestyle, time to rest and eat, family life, greater stress, greater access to processed or industrialized food, with processes that entail great environmental impacts on water, soil, and air.

With the advent of science and biotechnology, such as the production of high fructose corn syrup, coupled with the contradictory social policies on food that took place in governments since the 70's, where the increase in the consumption of soft drinks began. In the Mexican population, a disarticulation in achieving health indicators in the population laid the foundations since the boost to the sugar union that was justified with foreign companies with great conflicts of interest, and accelerated growth in the country, lead to a lack of control in terms of economy that will contribute to health promotion. It has always been said that Mexican food is extremely rich and diverse in ingredients, that is true, but it is also true that it contains many elements that are not suitable for a good diet. Knowing how we have developed our food culture allows us to understand the positive and negative effects it has had. The latter should lead us to reflect on the interactions that are generated with health, food, ecology, social, economic, and political aspects.

What effects does colonization have on a food system?

- Consider agroecological models instead of traditional ones: Faced with the health and environmental impacts that the prevailing agri-food system has generated, it is proposed as an alternative to agroecology, since it allows us to rebuild sustainable production systems that take as a principle the biophysical limits of nature, allowing biodiversity. Agroecology in turn promotes the channels between producers and consumers through access to healthy food in short proximity circuits, recognizing the importance of peasant wisdom and a right to food.

- Obesogenic environments have changed food culture and therefore the concept of health-disease.

- The loss and waste of food as a devaluation of food.

- Technology for social benefits and to increase the conservation of food without compromising the benefit of being consumed fresh to obtain the greatest benefit from its nutrients.

Finally, the exponential growth of the population leads to a demand for food that has justified the use of agrochemicals, but the production of more food also generates greater loss and waste of food, in addition to not achieving a nutritional enrichment of food in the face of soil. That has been deteriorating. Work life has led to both parents of a household having the need to both go out to work, without being able to invest in food education and therefore generate a food culture, leaving food and therefore destiny in the hands of the agribusiness of the health of minors or any member of the family. The lack of physical activity and time, coupled with greater access to energy-dense foods is what has contributed to the increase in the statistics of chronic degenerative diseases such as diabetes, hypertension, and dyslipidemias. Despite living in an era with greater access to information than ever before, the evidence shows that the act of eating and deciding is somewhat more complex, as is the complexity of human development itself, but this complexity is the methodology that It must be considered to address the current obesogenic environments through the approach and understanding of the meaning of food and health from an ecological and historical context.

Conflict of interest

The authors declare no conflict of interest.

Author details

Rebeca Monroy-Torres[1*], Graciela Velázquez-Delgado[2], Erika Carcaño-Valencia[3] and Gilber Vela-Gutiérrez[4]

1 Laboratory of Environmental Nutrition and Food Security, Health Sciences Division, University of Guanajuato, Mexico

2 Department of History, University of Guanajuato, Mexico

3 Social and Humanities Division, University of Guanajuato, Mexico

4 Laboratory of Research and Functional Products Development, Faculty of Nutrition Sciences and Food, University of Science and Arts of Chiapas, Mexico

*Address all correspondence to: rmonroy79@gmail.com

References

[1] FAO, IFAD, UNICEF, WFP and WHO. 2020. The State of Food Security and Nutrition in the World 2020. Transforming food systems for affordable healthy diets. Rome, FAO. http://www.fao.org/documents/card/en/c/ca9692en

[2] ENSANUT 2018. Informe de Resultados de la Encuesta Nacional de Salud y Nutrición. 2018. Available in: https://ensanut.insp.mx/encuestas/ensanut2018/informes.php

[3] Galanakis CM. The food systems in the era of the coronavirus (COVID-19) pandemic crisis. Foods. 2020; 523 (9): 1-10. Disponible en: doi:10.3390/foods9040523

[4] Niles MT, Bertmann F, Belarmino EH, Wentworth T, Biehl E, Neff R. 2020. The Early Food Insecurity Impacts of COVID-19. Nutrients. 2020; 12(7): 2096. Disponible en: doi:10.3390/nu12072096

[5] Ley de Aguas Nacionales. Última Reforma DOF 06-01-2020. http://www.diputados.gob.mx/LeyesBiblio/pdf/16_060120.pdf

[6] Food and Agriculture Organization of the United Nations. Guía para la transición hacia sistemas alimentarios y agrícolas sostenibles. 2018. Disponible en: http://www.fao.org/family-farming/detail/es/c/1147773/

[7] Jabbonsky, Larry. The Mexican Resurrection. Beverage World. 1993; 112 (1547): 38-39.

[8] Delgado-Ramos GD. Apropiación de agua, medio ambiente y obesidad. Los impactos del negocio de bebidas embotelladas en México. 2015. Editorial CEIICH-UNAM. Disponible en: http://www.giandelgado.net/p/libros.html

[9] Ruz-Barrio M A. Las huellas del ganado en el valle de Matlatzinco en el siglo XVI a través de los mapas hispanoindígenas: Relaciones Estudios de Historia y Sociedad. 2019; 35-72.

[10] Velázquez-Delgado G, Ayala-Calderón J. Mestizaje alimenticio de Guanajuato durante el Virreinato: Destiempos. 2014; 39: 13-26.

[11] Quiroz Enriqueta. Comer en la Nueva España. Privilegios y pesares en la sociedad en el siglo XVIII. Historia y Memoria. 2014; 19-58.

[12] Verdaderos preservativos del Chólera morbus, mandados publicar por el E. Ayuntamiento de esta capital. Guanajuato: Tipografía de Juan E. Oñate; 1850. 16 p.

[13] Victoria-Uribe M S. La minuta del día. Los tiempos de comida de la élite capitalina a principios del siglo XX: Historia y Grafía. 2010; 15-46.

[14] Meléndez J, Aboites-Aguilar L. Para una historia del cambio alimentario en México durante el siglo XX. El arribo del gas y la electricidad a la cocina. Revista de Historia Iberoamericana. 2015; 2: 76-101.

[15] Loría E, Salas E. Sobrepeso e integración económica en México. Economía Informa, 2014; 389: 3-18. [Consultado 2020 Dic 15] Disponible en: http://www.sciencedirect.com/science/article/pii/S0185084914721711 .

[16] Acuerdo Nacional para la Salud Alimentaria. Estrategia para el Sobrepeso y la Obesidad. 2010. https://www.paho.org/mex/index.php?option=com_docman&view=download&category_slug=vigilancia-prevencion-y-control-de-enfermedades&alias=385-acuerdo-nacional-para-la-salud-alimentaria-estrategia-para-el-sobrepeso-y-la-obesidad&Itemid=493

[17] Delgado, Gian Carlo, et al. La sustentabilidad en el siglo XXI. Interdisciplina. 2015. 3. (7). Pp. 9-21. SSN: 2395-969X.

[18] SEMARNAT. Informe de la situación del medio ambiente en México 2015. 2015. [Internet]. Available at: https://apps1.semarnat.gob.mx:8443/dgeia/informe15/tema/pdf/Informe15_completo.pdf

[19] CONABIO, 2020. [Internet]. Available at: https://www.biodiversidad.gob.mx/biodiversidad/porque

[20] SEMARNAT. 2017. [Internet]. Available at: https://apps1.semarnat.gob.mx:8443/dgeia/informe15/tema/pdf/Cap7_Residuos.pdf

[21] López, C. El estado del agua en México: retos, oportunidades y perspectivas. En: El agua en México. Actores, sectores y paradigmas para una transformación social-ecológica. Fundación Friedrich Ebert Stiftung. 2017.

[22] Rivera, et al. La obesidad en México. Estado de la política pública y recomendaciones para su prevención y control. Instituto Nacional de Salud Pública, UNAM, Academia Nacional de Medicina. 2018.

[23] Bejarano-González, F. (Coord.). Los plaguicidas altamente peligrosos en México. PNUD, UAEMEX, IPEN, INIFAP, UCCS, CIAD, RAPAM. 2017.

[24] Gobierno de México. Programa Nacional Hídrico 2014-2018. Diario Oficial de la Federación. 2013 Jun. [Consultado 2020 Dic 14] Disponible en: http://www.dof.gob.mx/nota_detalle_popup.php?codigo=5339732

[25] Gavito, M.E., Van der Wal, H., Aldasoro, E. M., Ayala-Orozco, B., Bullén, A.A., Cach-Pérez, M., Casas-Fernández, A., Fuentes, A., González-Esquivel, C., Jaramillo-López, P., Martínez, P., Masera-Cerruti, O., Pascual, F., Pérez-Salicrup, D. R., Robles, R. E., Ruiz-Mercado, I., Gilberto Villanueva, G. Ecología, tecnología e innovación para la sustentabilidad: retos y perspectivas en México. Rev. Mex. De la Biodiversidad. 2020. Available at: https://doi.org/10.1016/j.rmb.2017.09.001

[26] Oesterheld, M. Impacto de la agricultura sobre los ecosistemas. Fundamentos ecológicos y problemas más relevantes. Ecología Austral. 2018. 18:337-346.

[27] Dennis, C., Aguilera, J.M. Santin, M. Tecnologías que dan lugar al futuro: En: Agroindustria para el desarrollo. Eds: Da Silva, C.A., Baker, D., Shepherd, A.W., Jenane, C., Miranda Da Cruz, S. 2013.

[28] Suárez-Herrera, M.A, et al. Impacto de la ciencia y la tecnología en la salud y el cuidado del medio ambiente acuático. Biología. 2004. 18(1) OneFile: Informe Académico.

[29] Lorenzana, A. P. Seguridad alimentaria, tecnología y nutrición. Rev. Agroalimentaria. 2014. 8: 49-57.

[30] García-Casal, M.N. La alimentación del futuro: Nuevas tecnologías y su importancia en la nutrición de la población. Anales Venezolanos de Nutrición. 2007. 20(2): 108-114.

[31] Pérez V.A., Leyva T.D.A., Gómez M.F.C. Desafíos y propuestas para lograr la seguridad alimentaria hacia el año 2050. Rev. Mexicana de Ciencias Agrícolas. 2018.9(1): 175-189.

[32] RED-PDA 12.3. Estrategias para la Reducción y Pérdidas de Desperdicios de Alimentos. Día Mundial para la Concientización de las Pérdidas y Desperdicios de Alimentos. 2020.

[33] Monroy-Torres, R. Lo que la población debe conocer sobre los

resultados de la Encuesta Nacional de Salud y Nutrición (ENSANUT 2018). REDICINAySA. 2020. 4(1); 5-9. Available at: https://www.ugto.mx/redicinaysa/images/publicaciones/2020/REDICINAYSA-VOL-9-NO-1-ENERO-FEBRERO-2020.pdf

[34] Mullee A, Romaguera D, Pearson-Stuttard J, et al. Association Between Soft Drink Consumption and Mortality in 10 European Countries. JAMA Intern Med. 2019;179(11):1479-1490.